U0300581

当代建筑师系列

崔彤
CUI TONG

崔彤·建筑工作室　编著

中国建筑工业出版社

图书在版编目(CIP)数据

崔彤/崔彤·建筑工作室编著. —北京：中国建筑工业出版社，2013.10
(当代建筑师系列)
ISBN 978-7-112-15803-4

Ⅰ. ①崔… Ⅱ. ①崔… Ⅲ. ①建筑设计－作品集－中国－现代 Ⅳ. ① TU206

中国版本图书馆 CIP 数据核字（2014）第 039795 号

整体策划： 陆新之
责任编辑： 李 鸽 刘 丹
书籍设计： 康 羽
责任校对： 王雪竹 赵 颖

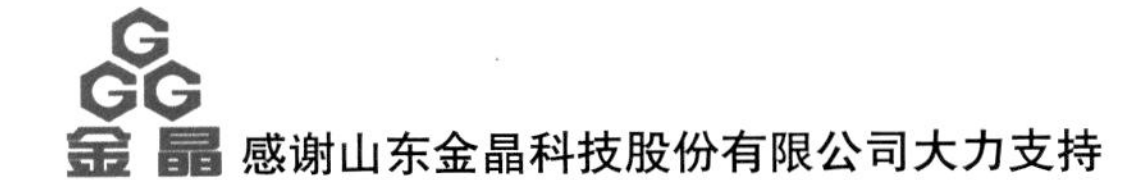

感谢山东金晶科技股份有限公司大力支持

当代建筑师系列
崔彤
崔彤·建筑工作室 编著
*
中国建筑工业出版社出版、发行（北京西郊百万庄）
各地新华书店、建筑书店经销
北京嘉泰利德公司制版
北京顺诚彩色印刷有限公司印刷
*
开本：965×1270毫米 1/16 印张：11¼ 字数：340 千字
2014 年 5 月第一版 2014 年 5 月第一次印刷
定价：98.00 元
ISBN 978-7-112-15803-4
(24546)

目 录 Contents

崔彤印象

文／黄元炤

崔彤，现为中国科学院建筑设计研究院的副院长、总建筑师，中国科学院研究生院建筑研究与设计中心创办人。他是中国建筑精神的捍卫者，致力于中国建构体系的重新阐释和实践，崇尚中国之道和现代之器的融合，这种交融的特殊性（中国建筑研究）已不言而喻地让他突出于云谲风诡的变局当中——时而中庸，时而纯净，时而理性，时而浪漫……尽管手法多样，风格多变，但给人既鲜明又深刻的中国精神的印记。

进入 21 世纪以来，“传统”与“现代”的争论依然持续着，各方存在不同的见解，但因为时代进步、经济全球化、大规模的建设发展等因素，使得我们似乎慢慢将“传统”遗忘或者刻意忽略……但崔彤并没有，他以一种“还原”的精神，从古老形态当中去找寻更原始的遗迹，为中国传统建筑找到根源所在——“架构”理论体系，这套体系是他发现与揭蔽的，源于人类最原始的居住形态，由巢筑在树杈上的“鸟居”生活状态演化成一种框架式的木构体系（穿斗与抬梁），是中国人独有的构筑方式。

中国科学院图书馆是崔彤于 21 世纪初的代表性作品之一，他尝试将中国精神在现代建筑中展现出来，从外在形式看，建筑暴露出许多巨大的梁柱与桁架，犹如是召唤着木构的形制，观察得出是探索“架构”体系后运用成形的，崔彤诠释为是一种“结构化的形式”，它统治着一个理性的内院和一个结构的建构体，开放而不包裹，真实而又传神。除此之外，这个项目也是崔彤个人“精神结构”的回响，反映的是他记忆中的一个精神场所的沉淀，图书馆就如同是儿时“家”的放大，与图书馆的图形关系同构的是一圈一圈环绕的就像家中的书架，仿佛一个巨大的环形四合院，他站在图书馆的中庭，向上看去，仿佛看到心中的“天顶画”那般。

“时空性”是崔彤关于中国建筑研究的另一项重要内容。无论是图书馆，还是中国工艺美术馆，或国家开发银行，在设计中，他试图创造一系列带有仪式性的节点与路径，传达一种叙述性的故事，引领人们去感受“空间的时间化”的作用，应用时空艺术的手法，将两种不同事件“剪辑”在一起产生一种戏剧式冲突。常规“城市中庭”、门洞、柱廊、台阶与广场，设立节点，延长路径，舒缓时间，细细品味，创造出空间的层次感，这是崔彤刻意营造的漫步式氛围，让人去“走”与“读”他的建筑，而这些高大进退的建筑物塑造出来的形式与空间，是倾向于“宏大化”、“巨大化”的“象征”设计语言的表述，为了体现一种国家精神，一种文化方面英雄主义情结的漫延，抒发公共知识分子对国家意识的一种至敬至高的情怀。

由于崔彤倾心于中国建筑体系的研究，将“现代”材料、技艺和“传统”的架构体系相结合，体现文化与技术的工艺精神。试图用木构的温暖和有机性去化解高技派的冰冷和机械感，并希望这个传承过去和跃向未来的“建构体”是源于场地和回应气候的新建筑。这个“建构体”在不同的场合中，有着不同身份，时而作为平衡体系，时而又作为生物体系。

在国家开发银行的设计中，崔彤是基于现象的透明性而确立将建构体作为平衡体系，期望在这个透明体系中同时处在新与旧、大与小的两组对立的关系系统中；在泰国曼谷的中国文化中心，崔彤敏感地发现热带地区菩提树具有原始建构体的特征，最终将“建构”转化为“种植”。实现所谓源于自然的木构又重新回到自然中的“生物体建构”。

国家开发银行与泰国曼谷中国文化中心是崔彤近几年来重要的作品，他的作品呈现出因势利导和复杂的融合，与其说是“折中”，毋宁说是“执中”。崔彤显然是用中国方法解决中国的问题，用中庸的“执两用中”对应于融和，即在辨析事物的两极之后寻求其“中”，在矛盾的对立面中寻求统一和平衡。而这个动态的平衡体系鲜明地脱胎于中国建筑的精、气、神，并一下子在时空转化中与现代建筑相遇，且希望产生一种突变。因此，他所坚持的“中国建筑精神”的设计主线没有改变过，烙着鲜明的情操，每件作品的成形就是一次设计总结，但其中也允许个别的不一样，崔彤认识到这是一项“悟道”的理性化过程，他希冀在宽广的设计生涯中找到那种属于知识分子的社会实践，在沉潜与内敛的态度上，用开阔的胸襟，将他的建筑在未来当中去熔化过去，这才是一种坚持，也是一种突破，更是一种“中国建筑”的超越。

Portrait

By Huang Yuanzhao

Cui Tong, the founder of Architectural Research and Design Center at Graduate University of Chinese Academy of Science, is currently the vice-president and chief architect of Architectural Design and Research Institute of Chinese Academy of Science. As a defender of the spirit of Chinese architecture, he devotes himself to the reinterpretation and practice of China's construction system and advocates the blending of Chinese philosophy with modern building theories. Despite how self-evident it should be, such one-of-a-kind integration (see "Research on Chinese Architecture")——at times moderate, or pure, or rational, or romantic has distinguished him from the ever-changing architectural domain. Therefore, although he designs with various skills and diversified styles, all his works distinctly and deeply bear the mark of Chinoiserie.

In the 21st century, the debate on "tradition" and "modernity" still prevails in every discipline, with each side holding different views. Due to such factors as social progress over time, global commercialization as well as large-scale construction and development, it seems that we have ignored "tradition" unwittingly or intentionally. However, Cui Tong, who does not follow suit and believes in the spirit of"back to basics", boldly steps into the historical forest to explore more primitive heritage buried in ancient architectural forms, trying to discover the spiritual fountain of Chinese traditional architecture——"frame-structure"theory system. This system, which Cui Tong founded and uncovered, originates from the most primitive inhabitation forms of human beings, that is to say, the lifestyle of "bird's nest" built on tree crotches evolving into a frame-like wooden structure system with column and tie beams, which is a building methodology unique to China.

As one of his representative works created in the beginning of the 21st century, the Library of Chinese Academy of Science is designed with the aspiration to showcase Chinese characteristics and cultural spirits within a modern structure. Seen from the outside, the building is featured with many exposed beams, columns and trusses which seem like a reminder of and a tribute to the ancient wooden structure. An elaborate observation will tell that such a design is the application of the "frame-structure" system exploration, which Cui Tong interprets into a "structuralized form". It governs a rational inner courtyard and a structuralized construction body, open and not enclosed, concrete yet highly expressive. Besides, the project also echoes to Cui Tong's personal "spiritual structure", mirroring a place for spiritual settlement in his memory. The library is the enlarged version of the house he once lived in as a boy; the encircling geometrical design is isomorphic to that of the bookshelves at home and resembles a gigantic traditional Chinese courtyard house. Standing on the atrium and looking upward, what rises before the eyes seems to be the "ceiling frescoes" in his heart.

"Time & Space" features are another subject of equal importance in Cui Tong's research on Chinese architecture. He is engaged to the creation of a series of nodes and paths with rituality and to the expression of narrative stories, leading people to experience the "temporalization of space" in his design works ranging from the library, China National Arts & Crafts Museum to China Development Bank. The Time & Space Art approach is employed to "edit" the two different events in one so as to address the dramatic conflicts. The design of conventional "city atriums", door openings, colonnades, steps and plazas combined with nodes as well as extended paths endows the space with multiple senses for people to savor slowly in a relaxing period of time. Visitors “wander” within the promenade atmosphere elaborately created by Cui Tong and harvest with distinct and personal interpretations. The forms and spaces shaped by these huge architectural elements pushed forward or drawn backward are the embodiment of a “symbolic” design language embracing “grandness” and “gigantism”. It is an expression of national spirit, a continuation and pervasion of heroism complex in cultural domain and the supreme emotion and sentiment of the intellectuals’ awareness about national consciousness.

Cui Tong, who is bent on the study of Chinese building system, combines the “modern” materials and techniques with “traditional” framework systems to demonstrate the craftsmanship of culture and technology. He aspires to offset the cold and mechanical feelings of high-tech buildings with the warm and organic characters of wooden structures, hoping that this “construction body” inherited from the past with a future vision can be a new architecture that is rooted in the setting and responds to the local climate. This “construction body” takes on multiple qualities under different circumstances, sometimes acting as balance system, sometimes as biological system.

In the design of China Development Bank, the construction body is taken as balance system on the basis of phenomenal transparency, expecting a compatible co-existence of the two contradictory relational systems of new and old as well as big and small within one same transparent system. Rather, as to the China Cultural Center in Bangkok, Thailand, Cui Tong is sensitive enough to notice the traits of primitive construction bodies that the tropic banyans bear, and transforms “constructing” into “planting”. Eventually, the wooden structure originating from nature returns to nature as a biological structure.

The China Development Bank and the China Cultural Center in Bangkok are two of Cui Tong’s significant projects in recent years. What is present in his works is the adaptability to different occasions and the intricate integration, which can be appropriately described as moderation rather than compromise. Evidently, Cui Tong is resolving China’s problems in manners that are peculiar to China. The Doctrine of Mean responds to the notion of blending, i.e. adopting what is in the “middle” after differentiating and analyzing the two ends of the matter, seeking for unity and balance between the two opposing sides of a contradiction. This dynamic balance system is a vigorous product of the essence, charm and spirit of Chinese architecture. It encounters with modern buildings as time and space evolves, and consequently induced an expected mutation. Therefore, he unwaveringly adheres to the “spirit of Chinese architecture”. Each of his works is a summary of his design philosophy with distinct imprint of personal sentiment despite a few exceptions. Cui Tong interprets it as a rational process of the pursuit for, meditation on and resonance to the truth and philosophical theories. Cui Tong anticipates to rightfully locate the social practices for intellectuals over the expansive design career. With a dedicated, low-profile and restrained attitude combined with an open mind, he fuses the past into the future in his works, which is as much a spirit of persistence and breakthrough as a transcendent achievement for Chinese architecture.

国家开发银行（复内 4–2 项目） 北京

China Development Bank (Project 4-2) Beijing

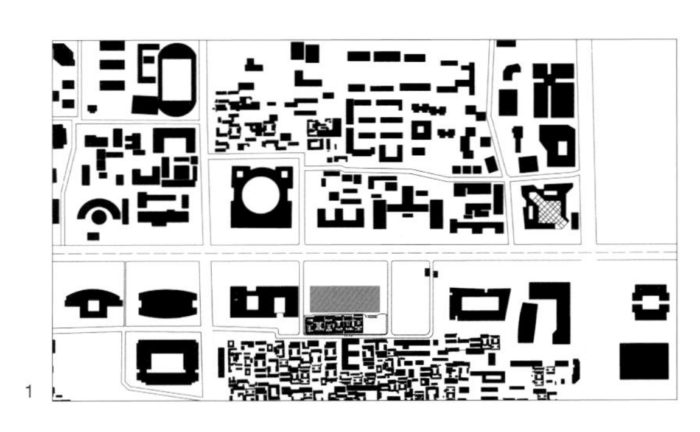

1 区位图
Location

2 建筑外景 1
Exterior 1

设计时间 / Design：2006
建成时间 / Completion：2013
建筑面积 / Building area：151000m²
项目组 / Design team：崔彤 赵正雄 何川 潘华 王欣 桂喆
合作 / Collaborator：北京市建筑设计研究院有限公司
业主 / Client：国家开发银行
摄影 / Photographer：杨超英

本案地处敏感的长安街西段，成为极具几何学精神的北京城中的重要建筑。总体布局源于对理性主义规划的巅峰之作——旧北京城历史文脉的重新思考和深入解析。

方案深入研究长安街周围环境及场所特征，以确定建筑所应具有的空间形态，空间秩序延续着长安街中统一的、均匀的类型学体系；建筑形态在传承如此永恒的、同构的、严谨的帝都威仪的同时，借用中国传统建造体系创造逻辑一致性的系统以回应城市。设计从传统街区的肌理研究出发，探究四合院的空间形态，寻找一种可能性，是在城市尺度与旧城肌理、城市空间与院落空间、现代技术与传统技艺之间建立一种平衡体系。

建筑的"双重性"源于现代与传统两类城市形态的并置与冲突，设计挑战是如何在两种异质要素之间寻找平衡，它要求建筑具有"宽容度"和"沟通"能力。架构体系的中国建筑具有空透骨架的独特品质，提供了一种与外界沟通的框架，建构的目的首先是要制造一个开放系统，它能够"连接"和"穿透"前、后、左、右而成为新媒介——它既不是满腹经纶的古典建筑，也不是"化装"后的殖民空间，它有力地根植于本土，"出生"在 21 世纪的北京。

设计过程伴随着从精神结构的建立到城市空间的建构。梁、柱作为建构体的基本单元存在于开放体系中，它们的组织关系优越于砌筑式的建筑。营造的合理性成为一种主题，集中体现在柱与梁、梁与顶的过渡关系中。"巨柱"体系在保留传统"柱"支撑的逻辑中演化为空间"柱筒"，9m × 9m 的"柱筒"空腔以求实现最小的空间单元。从巨"柱筒"封闭性的解体到"束柱筒"，从"束柱筒"又进一步形成"八柱七间"是基于传统秩序的建构。而超越结构体系的建构是在于空间的建构，这一新范式既区别于传统，也迥异于现代主义。不同于柯布西耶的均质柱网，也不同于密斯的无柱空间的均质化，"束柱筒"的介入实现了从"支撑单元"向分割空间的"空间单元"的转化。

四合院作为"第二级"空间单元在水平和垂直方向的"渗透"和"晕染"，形成具有同构关系的"八柱""七间"建筑，源于此时此地的巨构体系，将传统合院通过叠加、穿插、变异转化为立体的空中合院。在融于长安街建筑群体的同时，创造外表均质、内核节奏变化的新空间类型，成为庄严的银行总部办公楼。

This building is located in the western section of Chang'an Avenue, and is an important building in Beijing, which highly values geometry. The general layout is designed based on the reconsideration and in-depth study of the historical context of the capital city, which symbolizes the rationality in planning.

The design scheme made an in-depth study of the surrounding environment along Chang'an Avenue and the site conditions and it determined the space form of the building. Its space order follows the unified and well-distributed space system of buildings along Chang'an Avenue. The building's form inherits the eternal, isomorphic and rigorous dignity of Beijing, and at the same time, forms a logically unified system with the city by incorporating traditional Chinese architectural concepts. The design, which began by studying the patterns of traditional blocks and streets, explores the space form of traditional quadrangles. It aims to find a way to keep a balance between the city scale and the ancient city patterns, the city space and the quadrangles, as well as between modern and traditional techniques.

The "duality" of architecture results from the co-existence and conflicts between the modern and traditional city. Main challenges encountered by the design are to strike a balance between the two styles with different features, which requires the building to have a high degree of "tolerance" and "communication" ability. One unique characteristic of traditional Chinese architecture is that they have special structures, by which the architecture can communicate well with the outside environment. The primary goal of this construction is to form an open system that can provide a new intermediary to "connect" and "penetrate" different parts of the building. Therefore, the architecture is neither a typical traditional artwork nor simply a western style building with Chinese decor. Rather, it is a new design created in 21st century Beijing and is also deeply rooted in Chinese traditional culture.

The design process accompanies the whole process, from the establishment of the structure to the construction of urban space. The organizational relationship of the basic units such as beams and columns in pursuit of an open system is superior to that of a masonry-type construction. The appropriateness of construction becomes a theme, with a focus on the transition from columns to beams and beams to roofs. The "tremendous pillar" system evolves into "pillar casing", while retaining the logic of pillar support. In order to realize minimum space between units, the pillar cavity with the size of 9m × 9m shall be made hollow. The traditional order of construction starts from the disintegration of closeness of tremendous "pillar casings" to a "bundle of pillar casing", then to "eight pillars, seven rooms". The construction beyond the structural system lies in the special construction. This new paradigm differs from the traditional one and modernism. The interposition of a "bundle of pillar casing", is different from the homogeneous network of Le Corbusier and the uniformity of column-free space of Mies, achieves the conversion from "support unit" to the division of "space units".

2

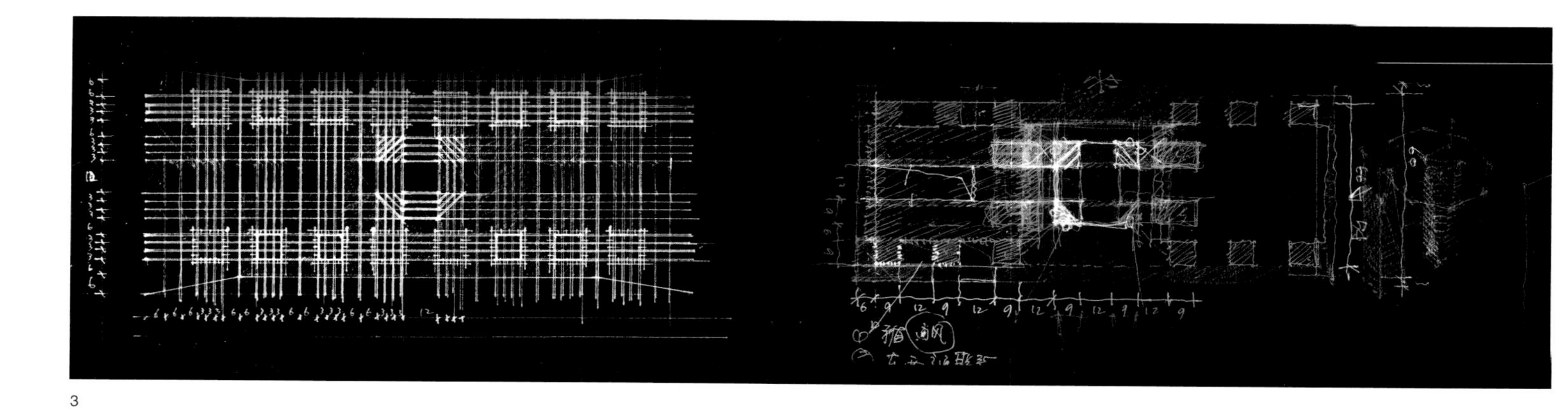

3

这个大尺度的建筑被当作一个“城市单元”而具有北京城特有的几何性和空间肌理。胡同、院落、广场等城市空间语言被移植到建筑中，平行于胡同和长安街的东西轴线串联两个边庭和城市中庭构成自身的秩序空间，边庭作为“第三级”空间单元将建筑一分为四构成4个单元，并提供阳光、空气。南北、东西轴线的交汇点上的“城市中庭”作为“现代北京”和“传统北京”的两种城市形态的过渡，以嵌入式的开放体系实现了建筑与城市环境的渗透关系，中央空透的城市中庭把旧城的“能量”和南向的阳光一同将长安街“照亮”。

设计延续了传统结构和美学的融合研究，对传统结构“原型”挖掘之后，探索一个新的建造系统，关注于建造逻辑、形式逻辑、空间逻辑有机互动的一套新体系。

巨构的形态是一个支撑系统，展现出多种空间的可能性，形成与银行总部相适宜的句法和语言，但巨构形态并没有游离于城市，相反以一种中国式的含蓄深藏于透明界面之后，构成被叠加之后的中国影像。

The quadrangles, as the "second level" space unit in the horizontal and vertical "infiltration" and "blooming", form the isomorphic construction of "eight pillars, seven rooms". The huge structure system is in concert with the traditional courtyard which is transformed into an aerial courtyard through superposition, interpenetration and variation. The new space genre of exterior uniformity and interior rhythmic change is created to become the headquarters for the bank while integrated into the group of buildings along Chang'an Avenue.

This large-size building is used as a "city unit" with a Beijing-specific geometry and spatial texture. The urban signs, like alleys, courtyards and squares, are transplanted into the building. The horizontal axis, parallel to the alleys and Chang'an Avenue, connects side halls and the city's halls to form its own orderly space. The side hall, named the "third-level" space unit, divides the building into four units and provides sunshine and air. The "city's halls" at the converging point on the horizontal and vertical axis stand for the transition between "modern Beijing" and "traditional Beijing", which define the infiltration of the building and urban environment through the embedded open system. The city's hall with a hollow center combines the "energy" of the old city and the sunshine facing the south to "light up" Chang'an Avenue.

The design acquires the fusion study of traditional structures and aesthetics, explores a new building system after studying the "prototype" of the traditional structure, and focuses on a new system with organic interaction between construction logic, formal logic and spatial logic.

The huge structure is a supporting system, displaying the possibility of many types of spaces. It forms a syntax and language which fits into the headquarters of the bank. But the huge structure is not disassociated from the city. Instead, it hides behind the transparent interface, symbolizing Chinese-style implicitness, and constitutes the Chinese image after superposition.

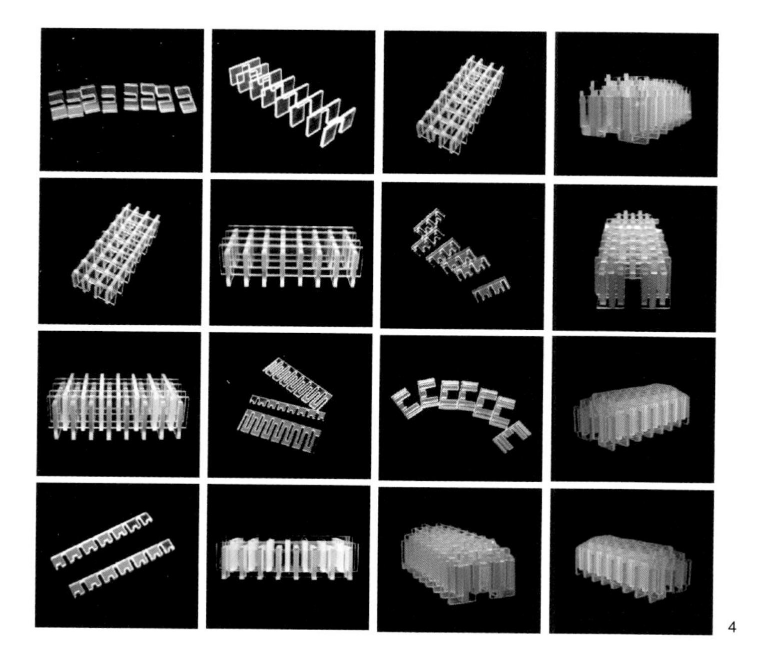

4

5

6

基地

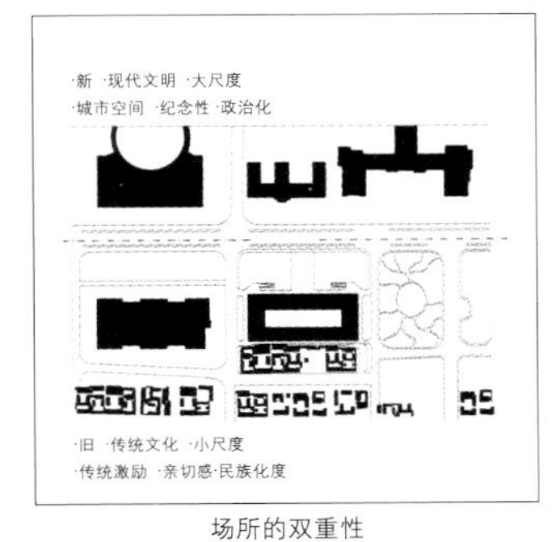

场所的双重性

平衡体系

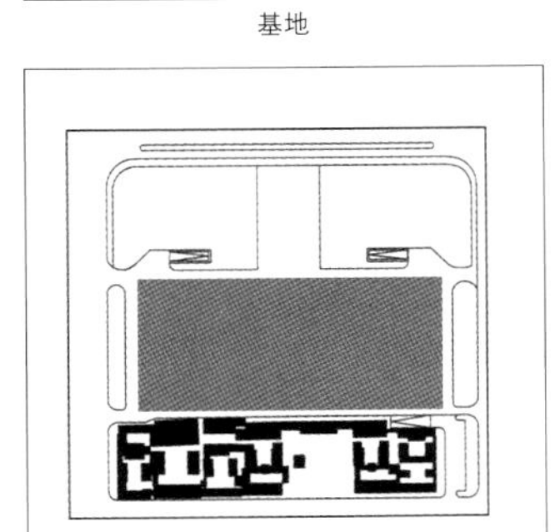

场所院落尺度

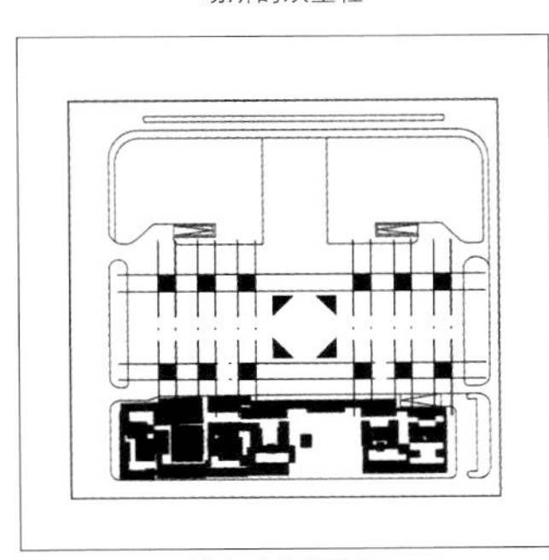

嵌入式空间秩序

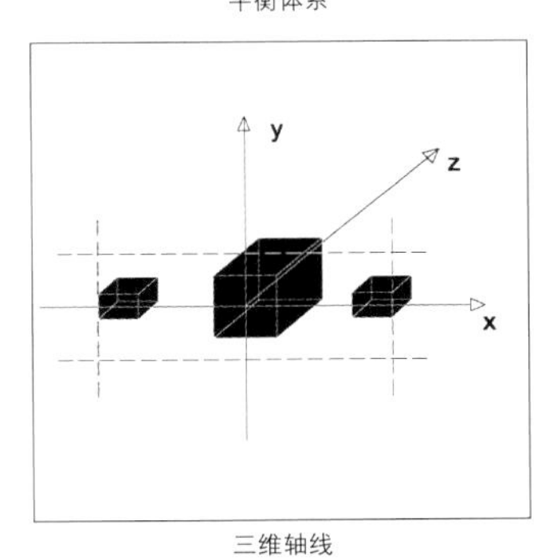

三维轴线

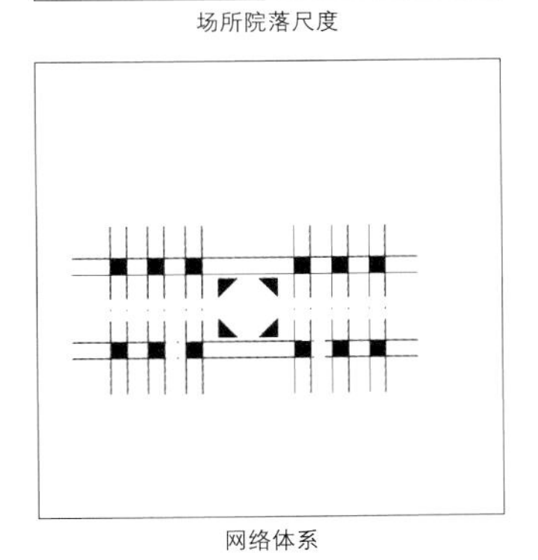

网络体系

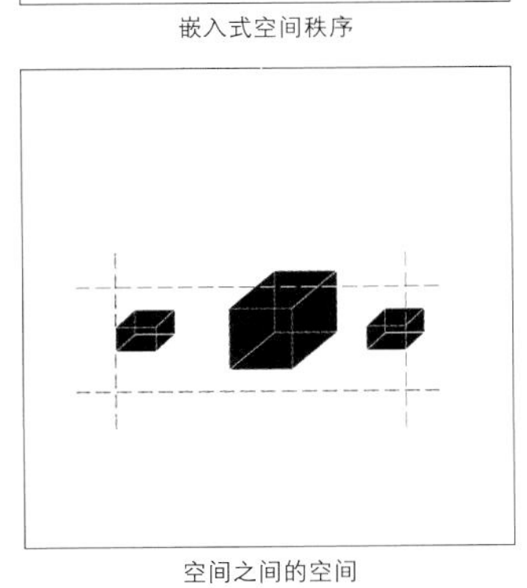

空间之间的空间

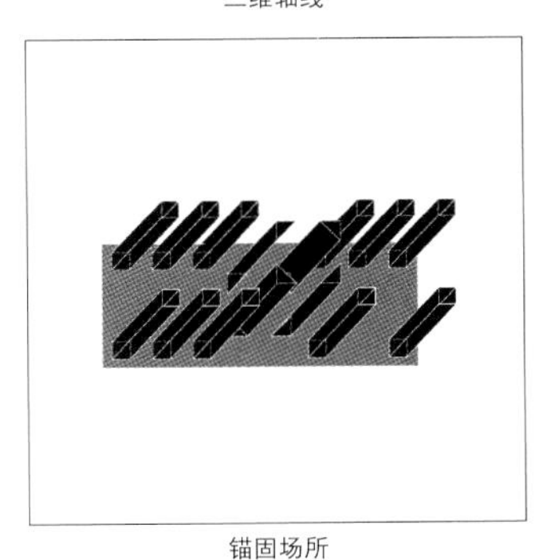

锚固场所

矩形架构

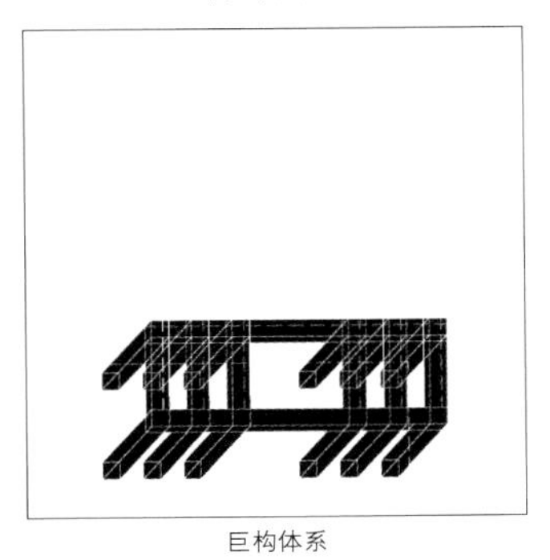

巨构体系

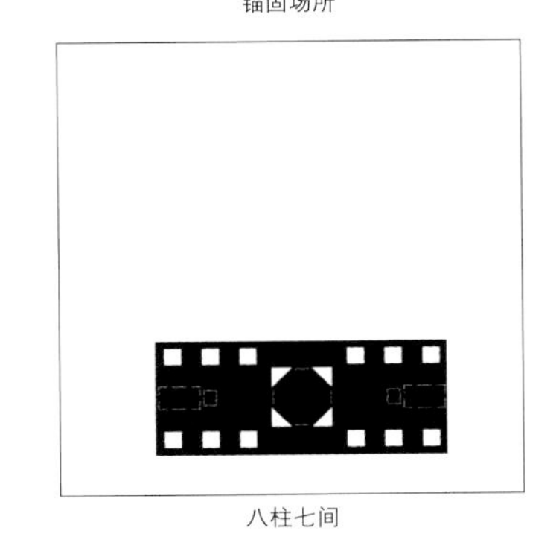

八柱七间

7

3 设计手稿 1
Hand sketch 1

4 模型 1
Model 1

5 模型 2
Model 2

6 设计手稿 2
Hand sketch 2

7 设计构思
Concept

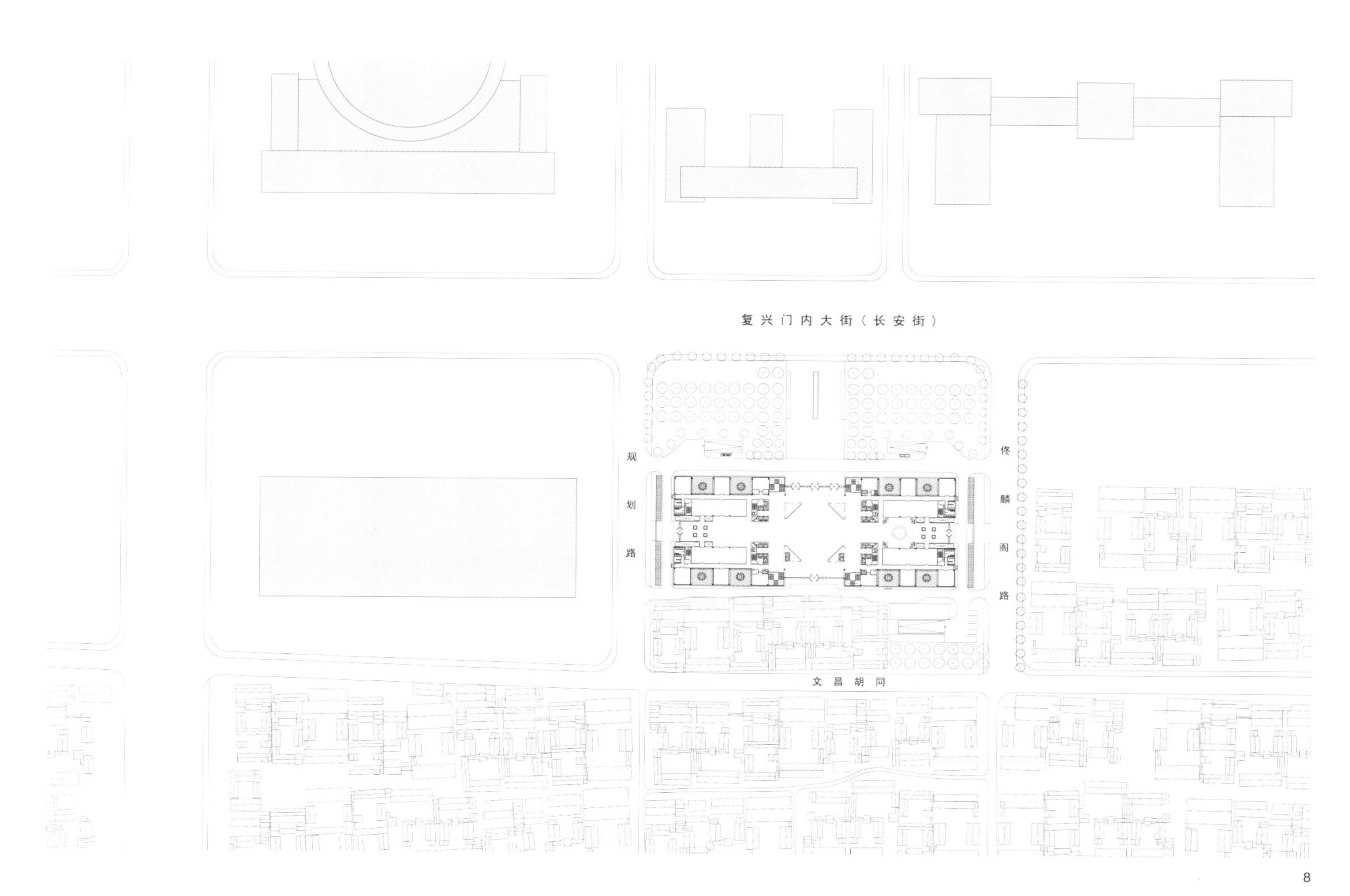

8

9

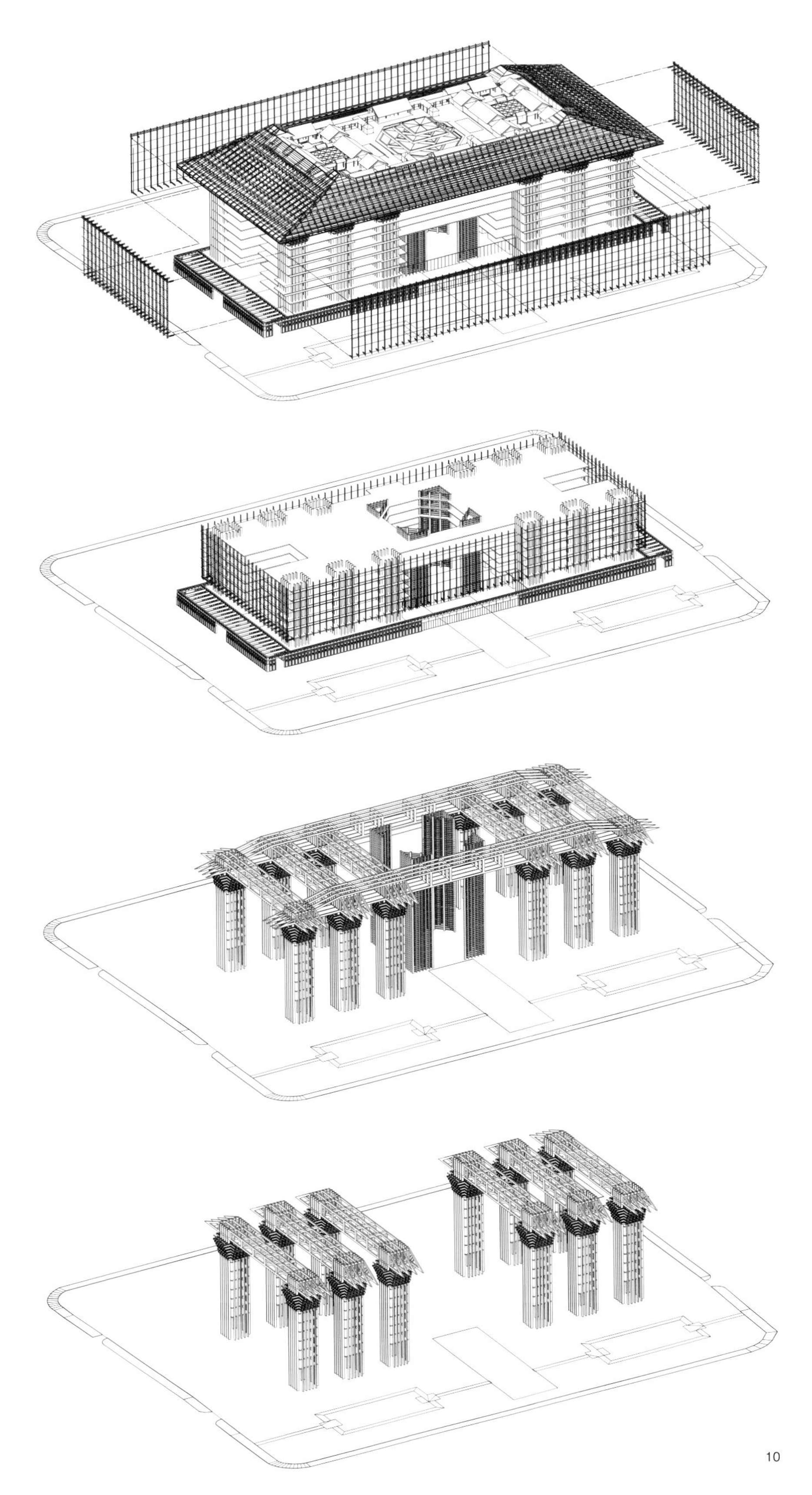

10

8 总平面图
Site plan

9 建筑外景 2
Exterior 2

10 解析模型
Analytic model

11 建筑外景 3
Exterior 3

11

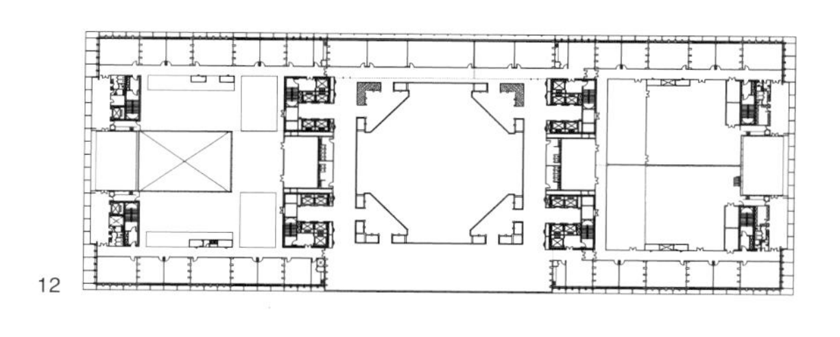
12

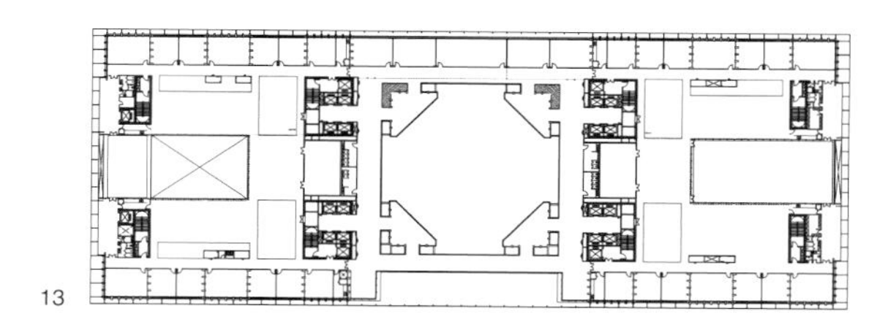
13

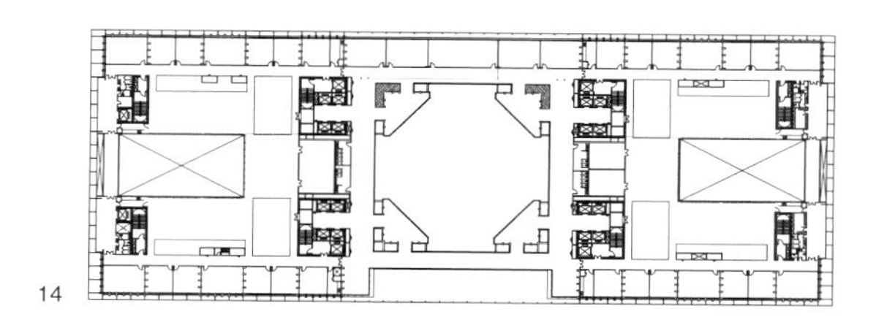
14

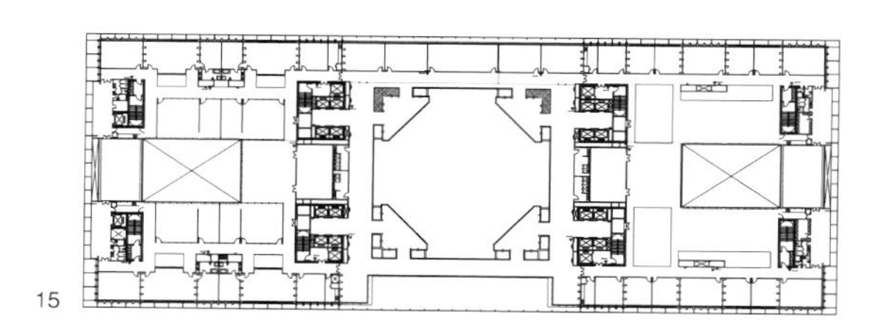
15

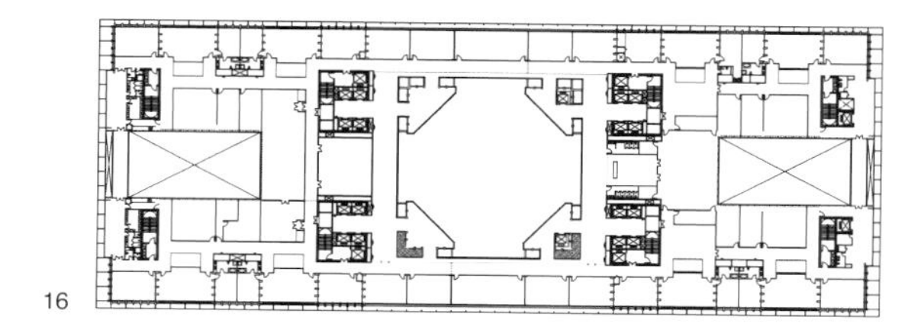
16

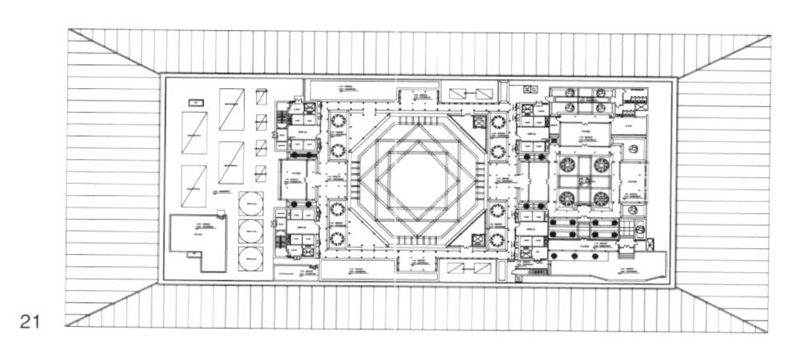
21

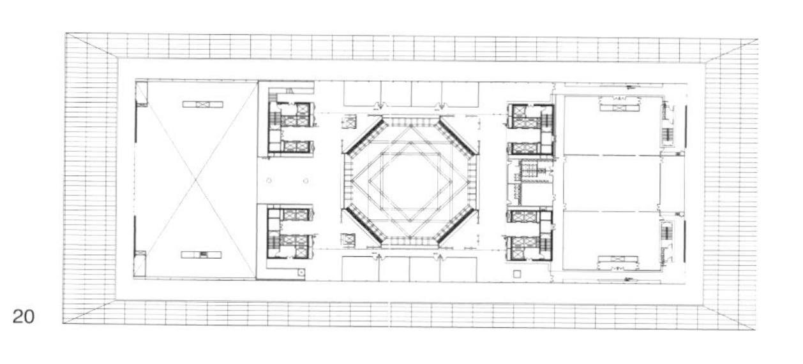
20

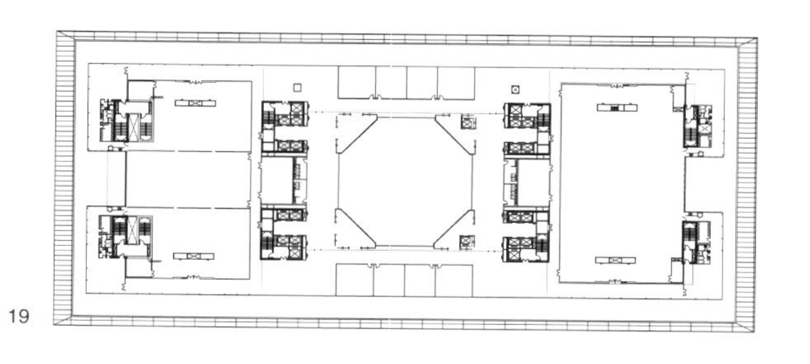
19

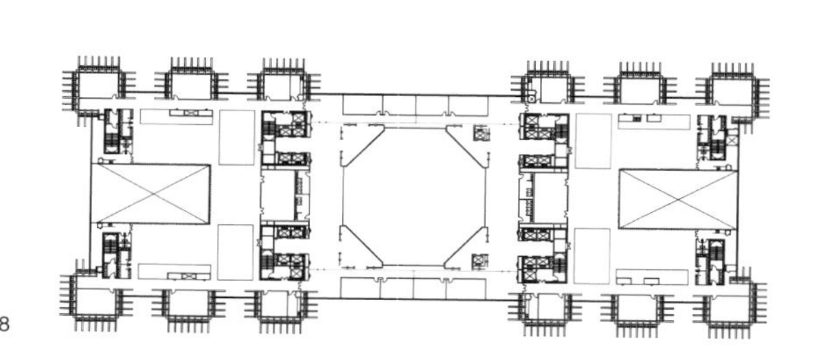
18

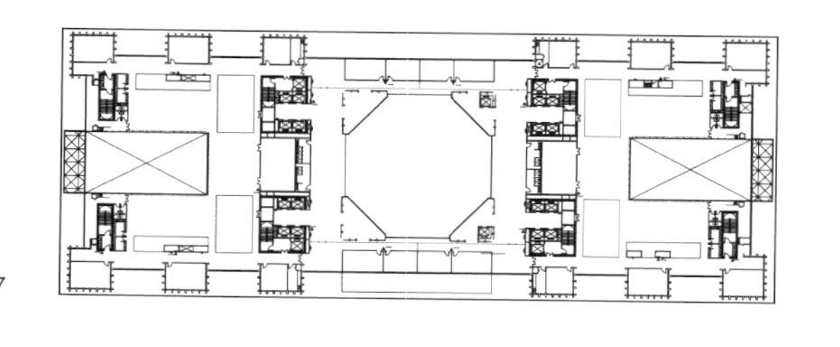
17

12 二层平面
The 2nd floor plan

13 三层平面
The 3rd floor plan

14 七层平面
The 7th floor plan

15 八层平面
The 8th floor plan

16 九层平面
The 9th floor plan

17 十层平面
The 10th floor plan

18 十一层平面
The 11th floor plan

19 十二层平面
The 12th floor plan

20 十三层平面
The 13th floor plan

21 屋顶平面
roof plan

22

23

24

23 建筑外景 4
Exterior 4

24 建筑外景 5
Exterior 5

25 建筑外景 6
Exterior 6

25

26

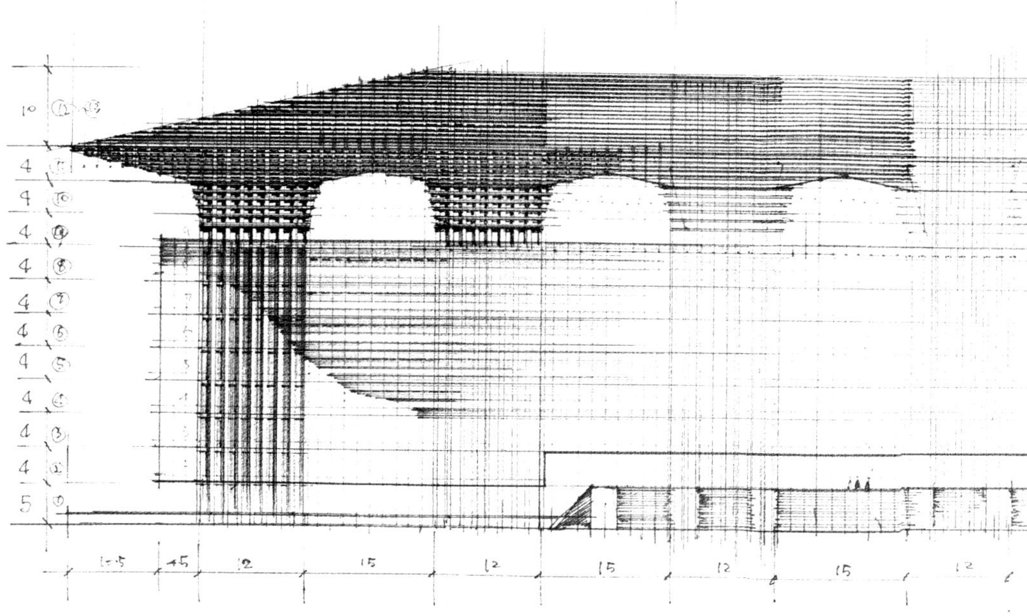

27

26 建筑外景 7
Exterior 7

27 设计手稿 3
Hand sketch 3

28 建筑外景 8
Exterior 8

28

泰国曼谷·中国文化中心　　泰国

The Chinese Cultural Center in Bangkok　Thailand

设计时间 / Design：2008
建成时间 / Completion：2012
建筑面积 / Building area：7650m²
项目组 / Design team：崔彤 王一钧 桂喆 吕僖 陈希 周军 房木生 苑蕾
合作 / Collaborator：Plan Architect co. Ltd
业主 / Client：中国文化部
摄影 / Photographer：刘崇明 范虹

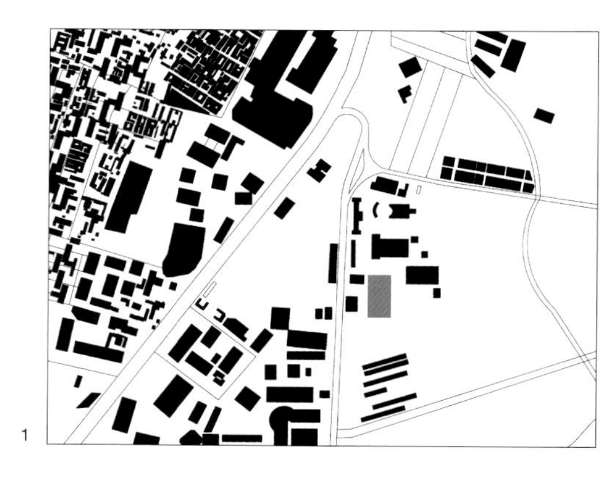

1 区位图
Location

2 建筑外景 1
Exterior 1

位于泰国曼谷的中国文化中心，由两组建筑单元错动连接成“Z”形体块，构成两个外部空间：一个外向型面向社会和民众的广场；一个内向型静谧的中国园林。建筑与外部空间的嵌套式的关系使之成为整体。庭院和广场与建筑内部空间的联系，通过内外空间不断地过渡与转化形成具有“东方时空”理念的场所。

“文化中心”作为一种特殊类型的外交空间，是中国文化传播和中泰文化交流的重要场所，它不可避免地要回答“中国化”“泰国性”等问题。尽管“图像式”“形式化”的语言是一种常用表征手段，但“标签”终归不能全面回答“中国化”问题。对于在异邦的中国文化中心首先体现在为“活动者”提供一个吸引人的、渗透着中国文化的探访空间，它既不应该是强加式的，也不应该是简单复制出来的，而是在特殊的土壤中被培养出来，并或多或少具有改良的特质，好像是中国的“种子”被移植到异国他乡存活后才显示出活力。文化中心的建构也同样基于“生物学”的生存方式，并对当地的气候、环境作出回应，在这一过程中，不可缺少的环节包括生长、适应、改良、变异。正如同佛教进入中国和泰国被改良为不“范式”的佛教，其基因的改变是自我生存机能的调节，以便得到进化和重生。因此文化中心的建构其实在于场所的重构，包含着适应环境、改造环境和表达环境，这一过程伴随着谨慎“优选”传统文化的基因，在地脉与文脉的培养中，促进一种交融的文化。

建筑形态通过水平密檐寻求中国古典建筑的相关性，正面的中国建筑形态特征体现在水平向的延展，侧面关注垂直向度上的重叠，寻找与泰国寺庙建筑的相关性。而这一形态的根本出发点是对当地湿热气候的回应。

The Chinese Cultural Center in Bangkok is a Z-shape construction made by the dislocation of two units. Therefore, it has two exterior spaces：an external plaza that is open to the public, and an internal tranquil Chinese garden. The building and the exterior spaces are well positioned, forming an integrated whole. The connection between the interior space, the courtyard, and the plaza is realized by the constant transition from the interior and the exterior, thus creating a place with a typical Asian style.

As a special space with diplomatic functions, the cultural center is an important place to spread Chinese culture and for Sino-Thai communication. Therefore, it is inevitable in its function to answer questions about China and Thailand. Although the language of "image type" and "formalization" are common means of representation, labelscan never fully answer questions about China. As a Chinese cultural center located in the foreign country, first of all, it shall provide visitors with an attractive visiting place richly permeating with Chinese culture. It should not be imposed or simply copied; instead it shall be cultivated in special soil and can be more or less modified. It seems that only when transplanted to grow in a foreign land, China's seeds can show their vitality.The construction of the cultural center is also based on the survival mode of "biology", and so it must respond to the local weather and environment. This process includes four indispensable steps：growth, adaptation, modification and variation. Just as Buddhism was modified to be non-paradigm after entering China and Thailand, the change of its genes means the adaptation of its self-survival functions and then evolution and rebirth. Therefore, the building of the cultural center lies in the construction of the site, encompassing adaptation to the environment, transformation of the environment and reflection of

2

泰国曼谷属低纬度热带气候，特殊环境也孕育了特殊的种群和文化。作为建筑的基本设计架构，"防雨"、"遮阳"、"通风"，其实早已存在于林木之中。我们的设计程序是观察、发现，并选取最具生命特征的自然建构体；设计的方法论源于自然秩序而发展至辉煌的中国木构体系，重新还原给自然，在这个重构空间的过程中中国式的建构体系在"进化"，仿佛于热带丛林中的造物，架构的空灵、悬挑技艺、生长逻辑，在这片温润的地脉中衍生出一股东方的豪劲。

the environment. In this process, the genes of the traditional culture are cautiously "prioritized"; during the cultivation of the land and culture, an integrated culture is promoted.

The architectural form seeks the relevance of China's classical architecture by horizontal tight tiles. The front facade of Chinese architectural form displays horizontal extensions; the lateral side concentrates on vertical superposition, pursuing a similarity with the architectural form of Thai temples. The primary concern for this form is to adapt to the local weather, which is humid hot.

Thailand, located at a low altitude, is in a tropical climate zone. This unique land gives birth to special species and culture. The fundamental design concept for the building – waterproofing, shade-providing and ventilating– has existed in the forest for a long time. Our design procedures include observation, discovery and selection of natural construction forms which show the strongest vitality. The design methodology is derived from natural order, developing into a glorious Chinese timber system, and then restores itself to the nature. In this process of space reconstruction, the Chinese-style construction system evolves. It is as if the creatures, elusive buildings, and cantilever techniques in the tropical forest are generating Asian vigor in the enriched land.

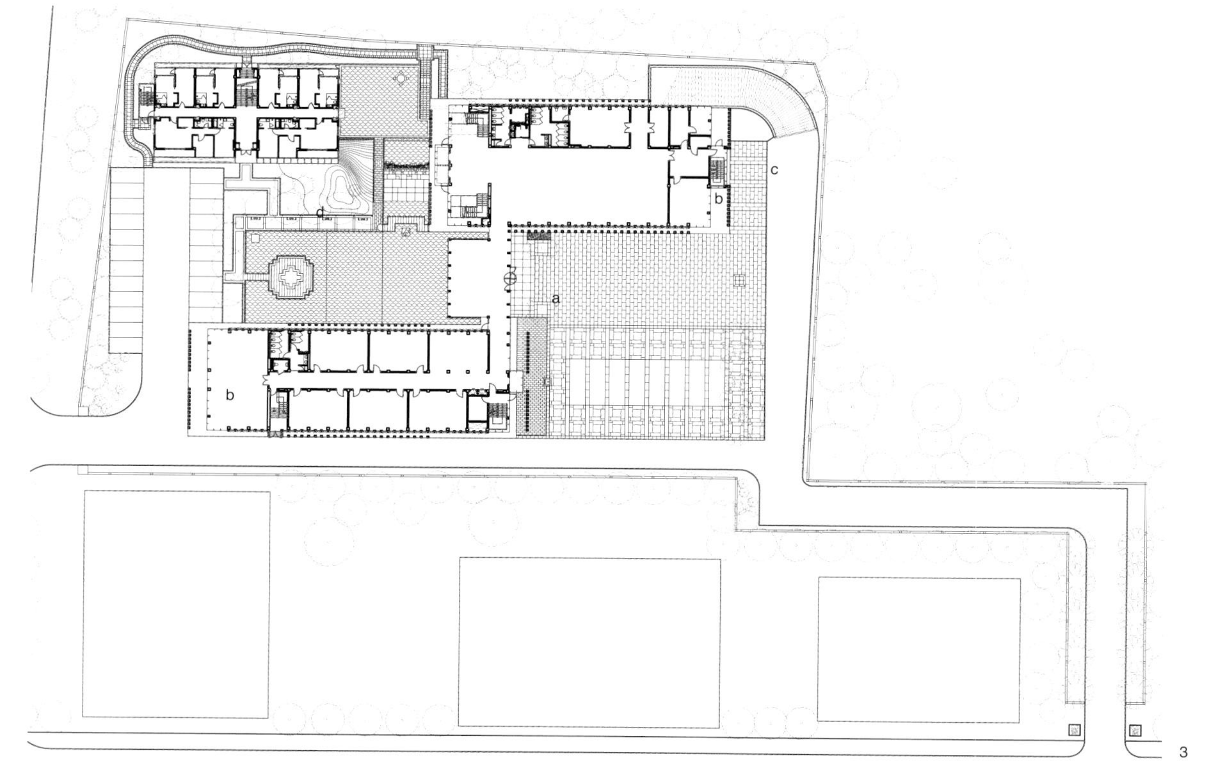

3 总平面图
Site plan

a 建筑主入口
Main entrance
b 建筑次入口
Auxiliary entrance
c 车库入口
Garage entrance
d 建筑庭院
Courtyard

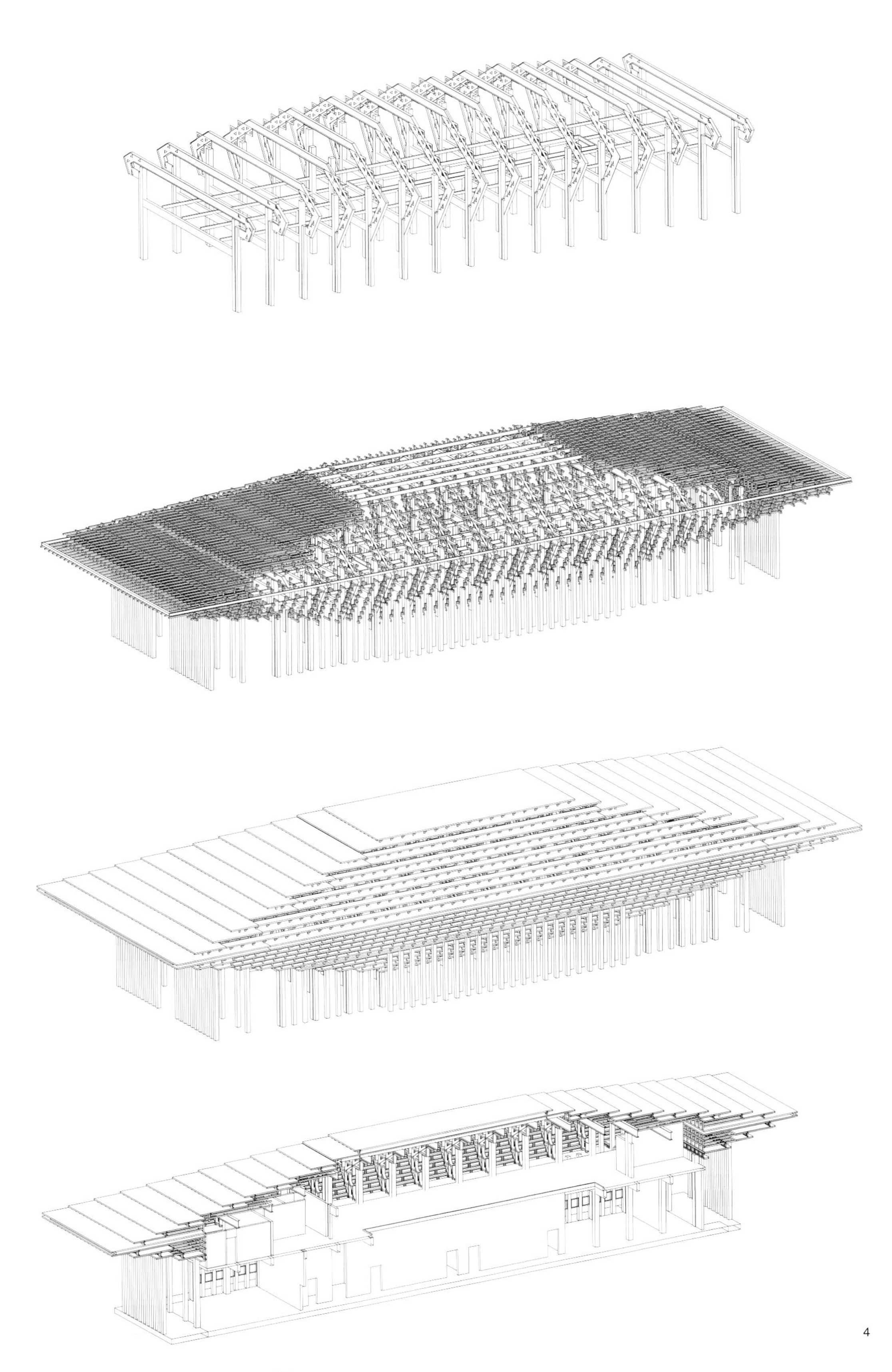

4 解析模型
Analytic model

5

5 建筑外景 2
Exterior 2

6

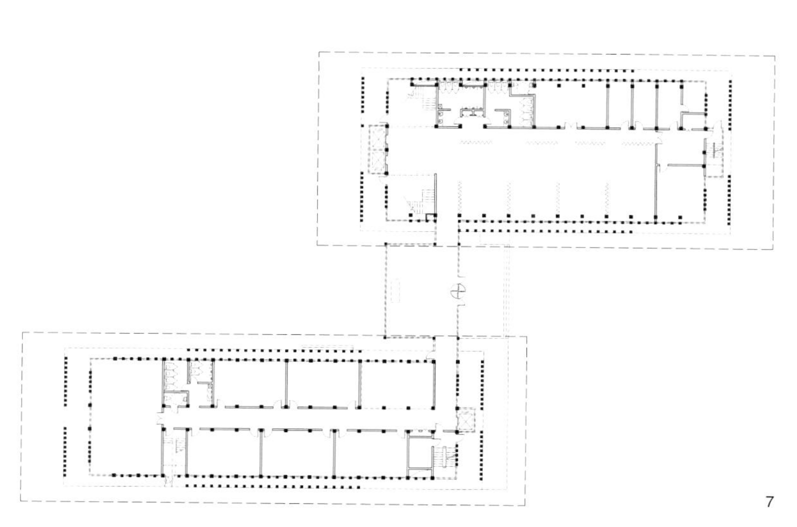

7

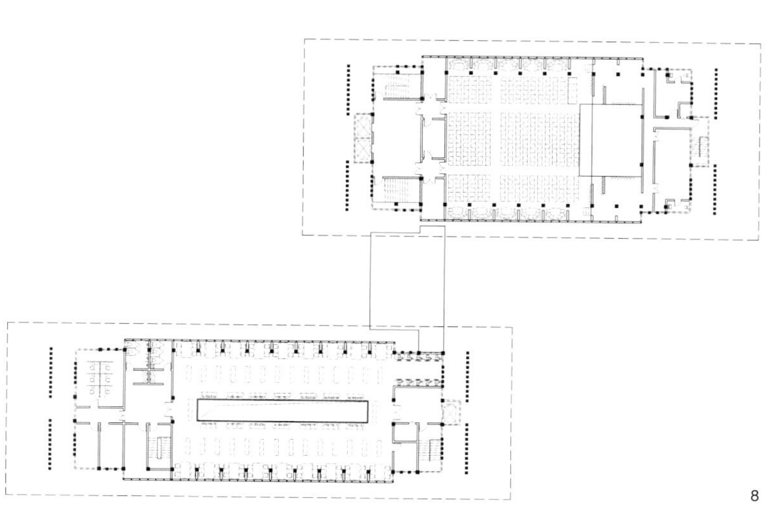

8

6 建筑外景 3
Exterior 3

7 首层平面图
The 1st floor

8 二层平面图
The 2nd floor

9 建造过程
Construction process

9

10

11

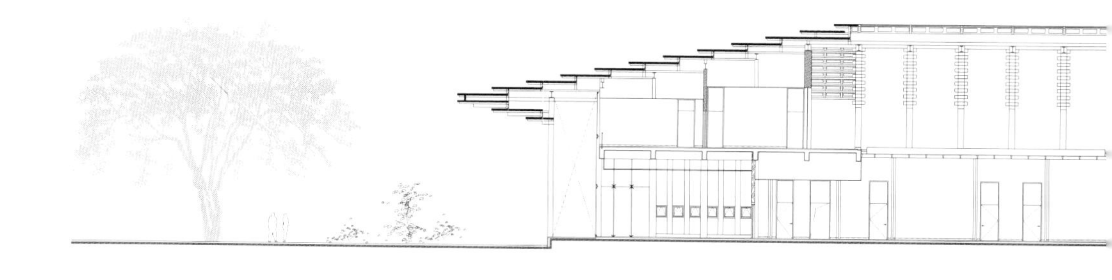

10 建筑庭院 1
Courtyard 1

11 建筑庭院 2
Courtyard 2

12 建筑庭院 3
Courtyard 3

13 剖立面图
Section & elevation

12

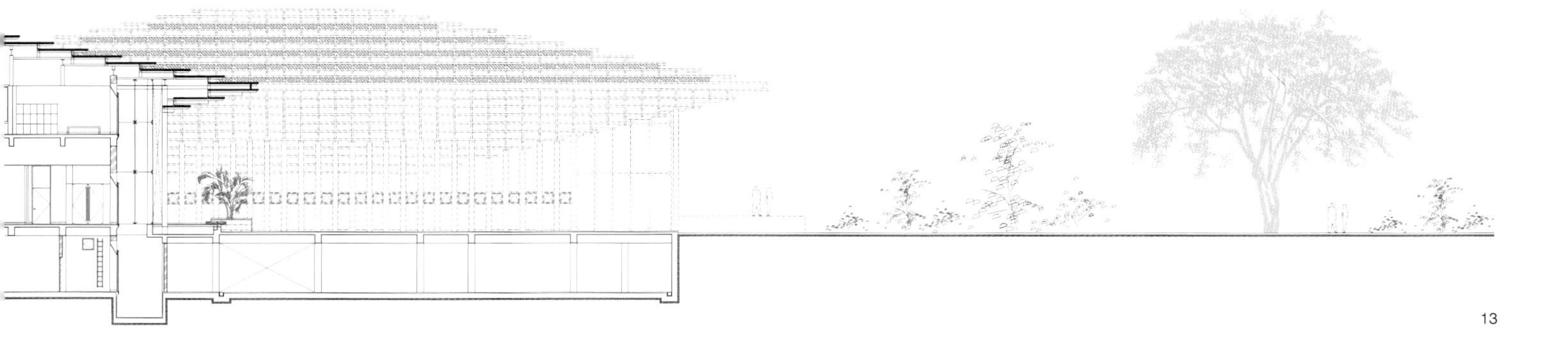
13

14

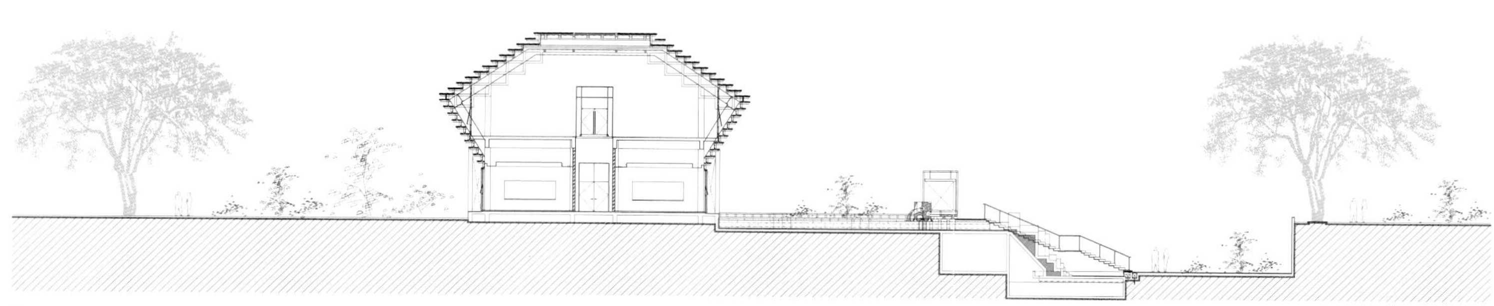
15

16

17

18

14 建筑室内 1
Interior 1

15 剖面图
Section

16 建筑室内 2
Interior 2

17 建筑室内 3
Interior 3

18 建筑室内 4
Interior 4

19

19 建筑外景 4
Exterior 4

20 南立面图
South elevation

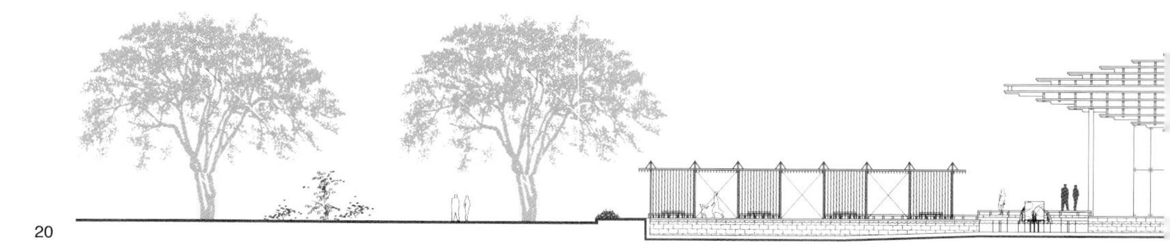

20

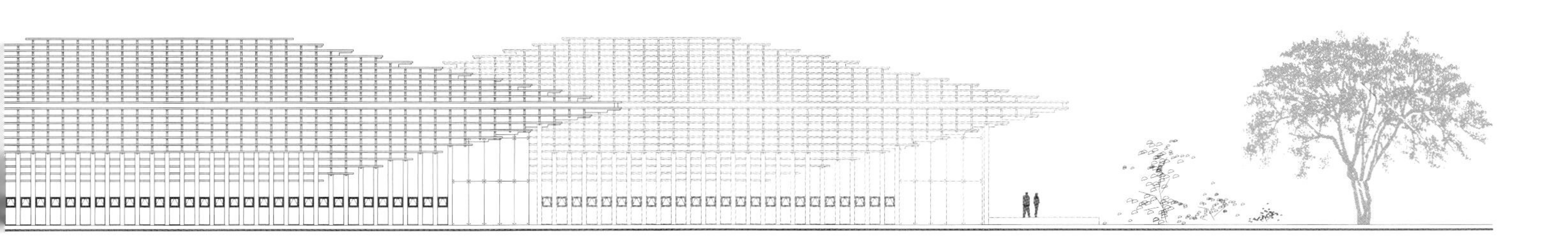

中国工艺美术馆　北京

National Arts & Crafts Museum of China　Beijing

设计时间 / Design：2011 ~ 2012
建成时间 / Completion：N/ A
建筑面积 / Building area：86800m²
项目组 / Design team：崔彤 刘建平 赵迎 王一钧 张润欣 王康 陈希 司亚琨 曹炜 房木生 周军 许楠
业主 / Client：中国文化部　中国艺术研究院

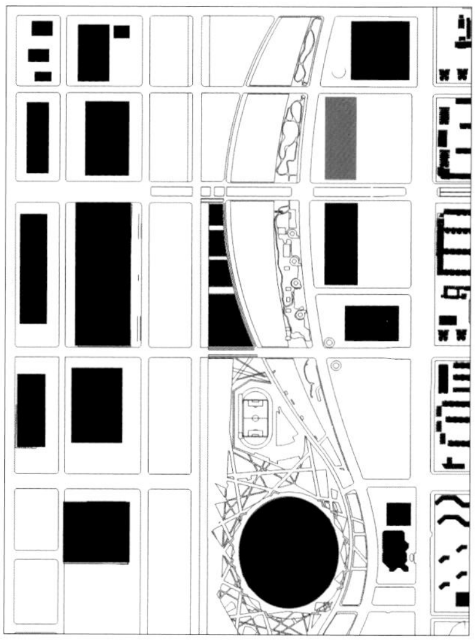

作为非物质文化的中国木构建筑，几千年的营造达到了非常精湛的程度。中国木构体系在技术的追求中，又融入了艺术表达，并形成“技艺”合一的建构体系。中国工艺美术馆和非物质文化遗产馆，以“架构式”的营造技艺建构出工艺精湛的壮美建筑。在这个巨型的工艺品中凝聚着建筑、桥梁、船只、中型器物、小型饰物共同的设计“意匠”，以“大设计系统”融合为一种“精神结构”统摄全局、一生万物。

传统的架构体系是梁柱作为基本构件不断生成结果。梁柱的“二木相合”的建构技艺转化为一种抽象的连接关系，以二元要素的相关性衍生出多种建构形态：砌筑→架构→编织。两种构件的叠置、砌筑、建成的坚固有力壁体；两种构件的穿插、咬合架构出空灵的骨架体系；两种构件的穿梭编织出半透明丝织般的网纹肌理，光的纵横交织构成轻盈和诗意的建构形态。将梁柱所形成结构体系多“功能化”，它不仅扮演结构支撑体系的角色，也参与到诸如空间、形态、功能和装饰层面中，建筑成为有机的一体化系统，如同中国传统建筑结构美、形式美、装饰美有机结合为一体，结构等于形式。

源于北京中轴线北奥运公园自然山水意象，源于工艺美术精湛技艺和非物质文化遗产的动态文化传承，建筑成为一种对环境作出反应的有机体，建筑仿佛源于自然秩序而生成。

The wooden architecture, as the Chinese intangible culture, has reached a very exquisite degree in thousands of years. In the pursuit of technology, the Chinese also focus the art expression, they have created a perfect construction system which was integrated the skill & art. By this construction system, China National Arts and Crafts Museum is constructed as a magnificent building.

Traditional construction system is based on the girder and column. It can be abstracted as a simple connection of "two wood". This simple connection has derived lots of other construction form. They can be built as a strong wall, a dense structure or transparent texture. China National Arts and Crafts Museum was combined structure, form and decoration into a whole by this traditional construction system. The structure is the form.

1 区位图
Location

2 建筑细部
Exterior Details

2

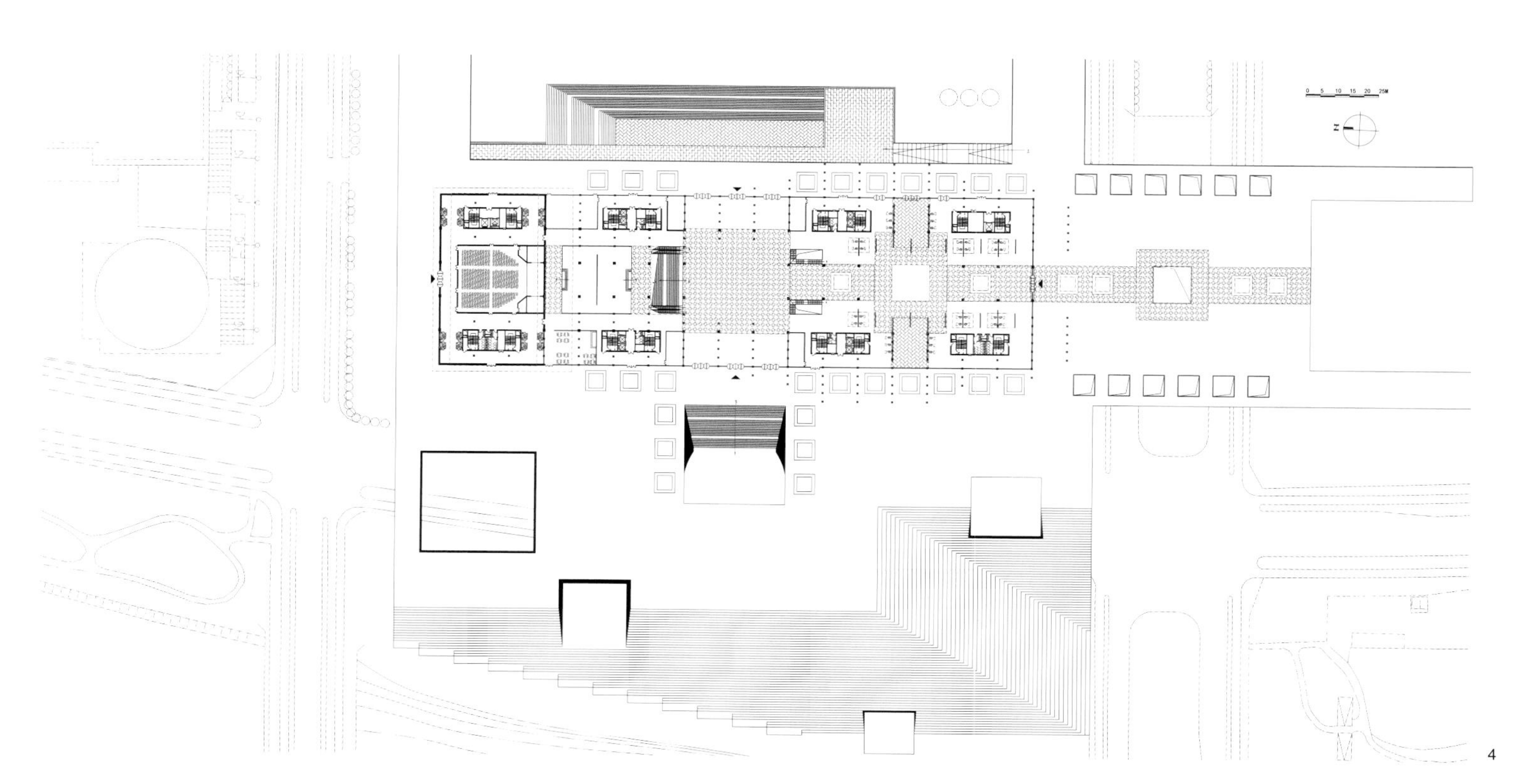

4

3

3 建筑外景 1
Exterior 1

4 总平面图
Site plan

5 建筑解析
Analysis

5

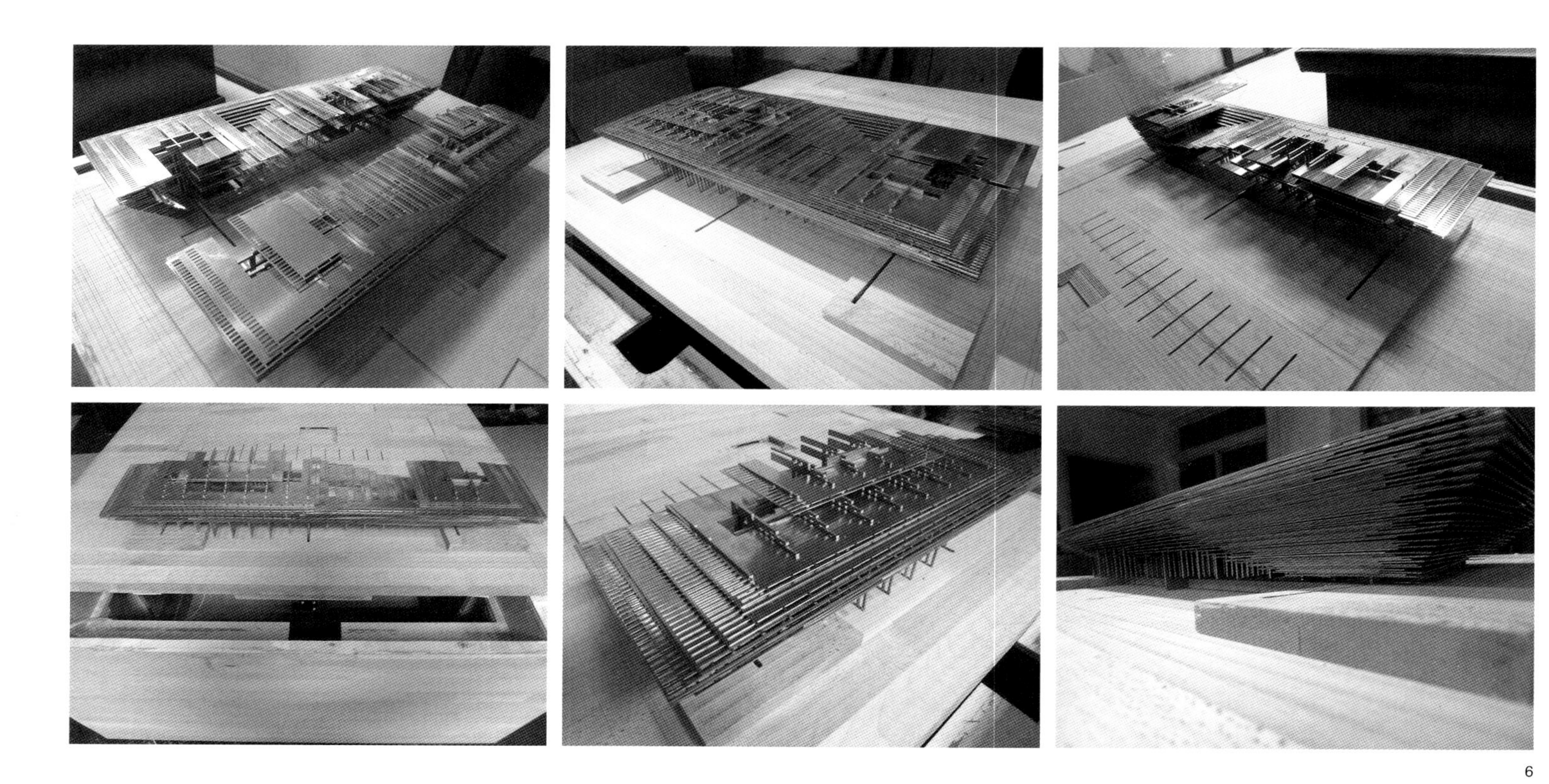

6

6 建筑模型
Model

7 建筑外景 2
Exterior 2

7

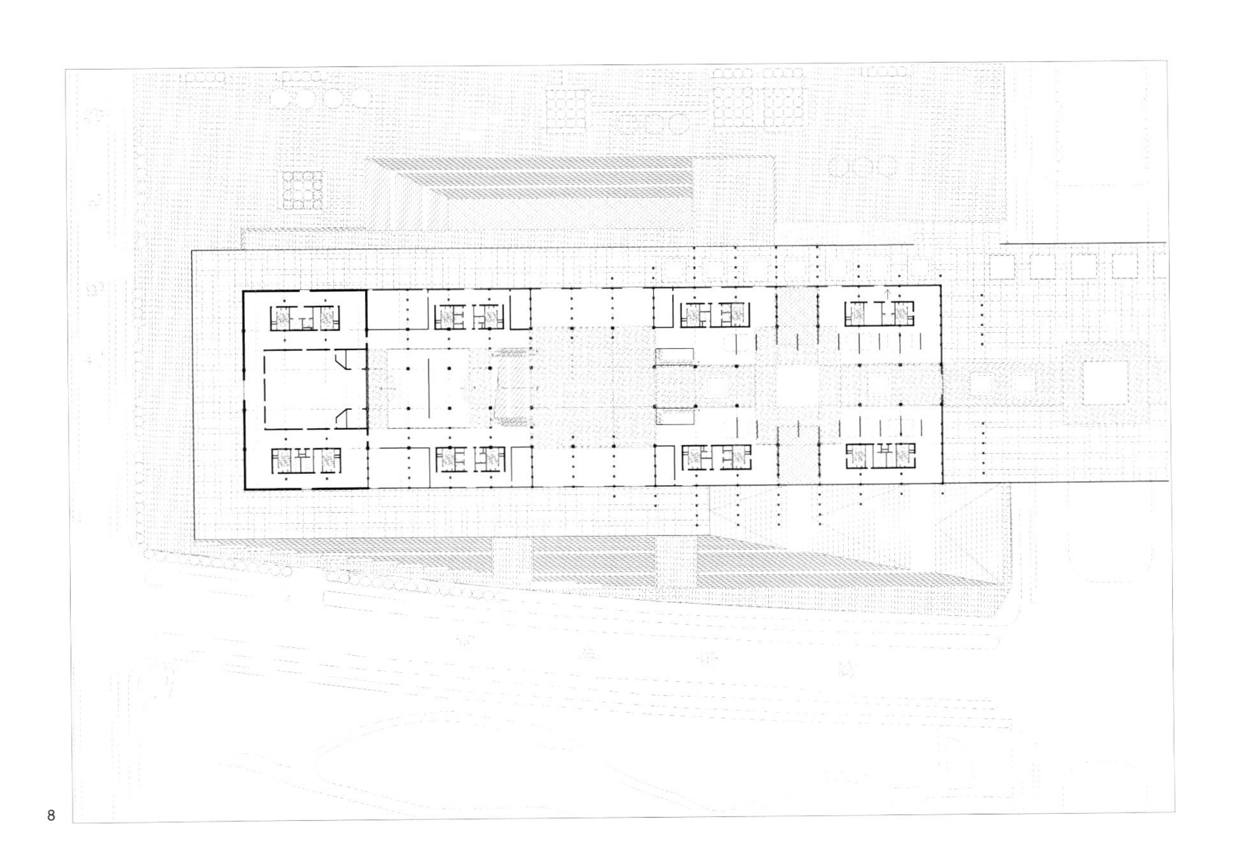

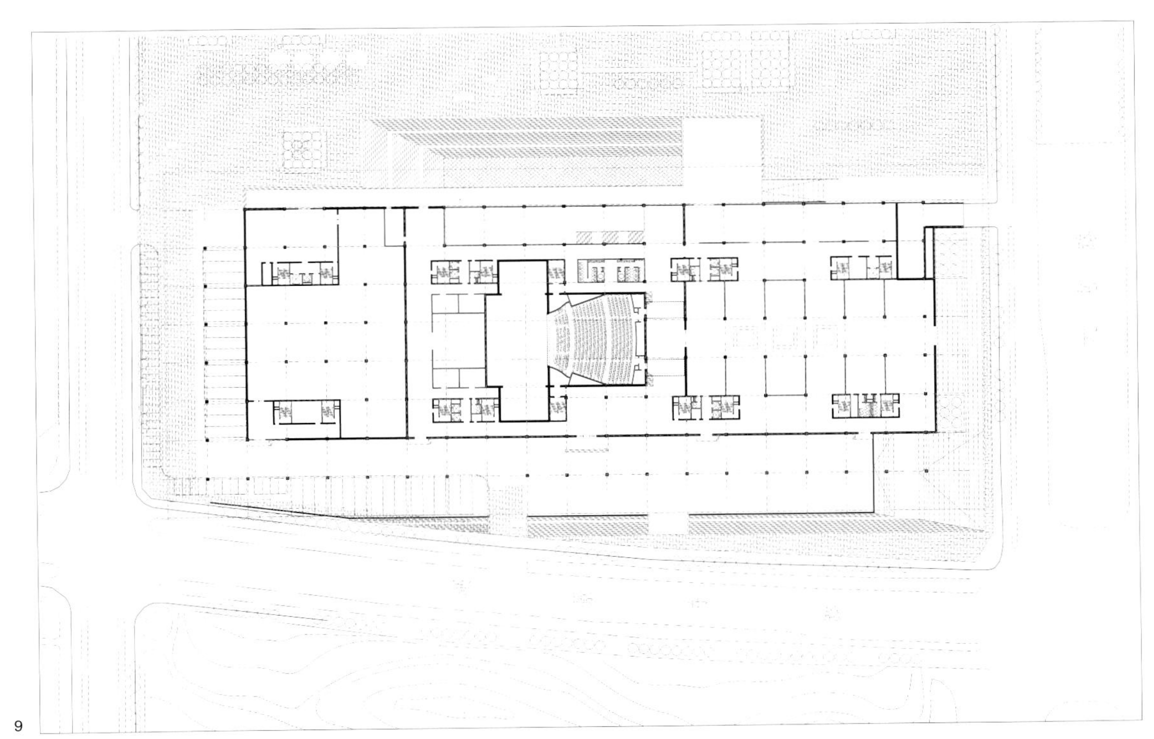

8 首层平面
The 1st floor plan

9 平台层平面
Platform floor plan

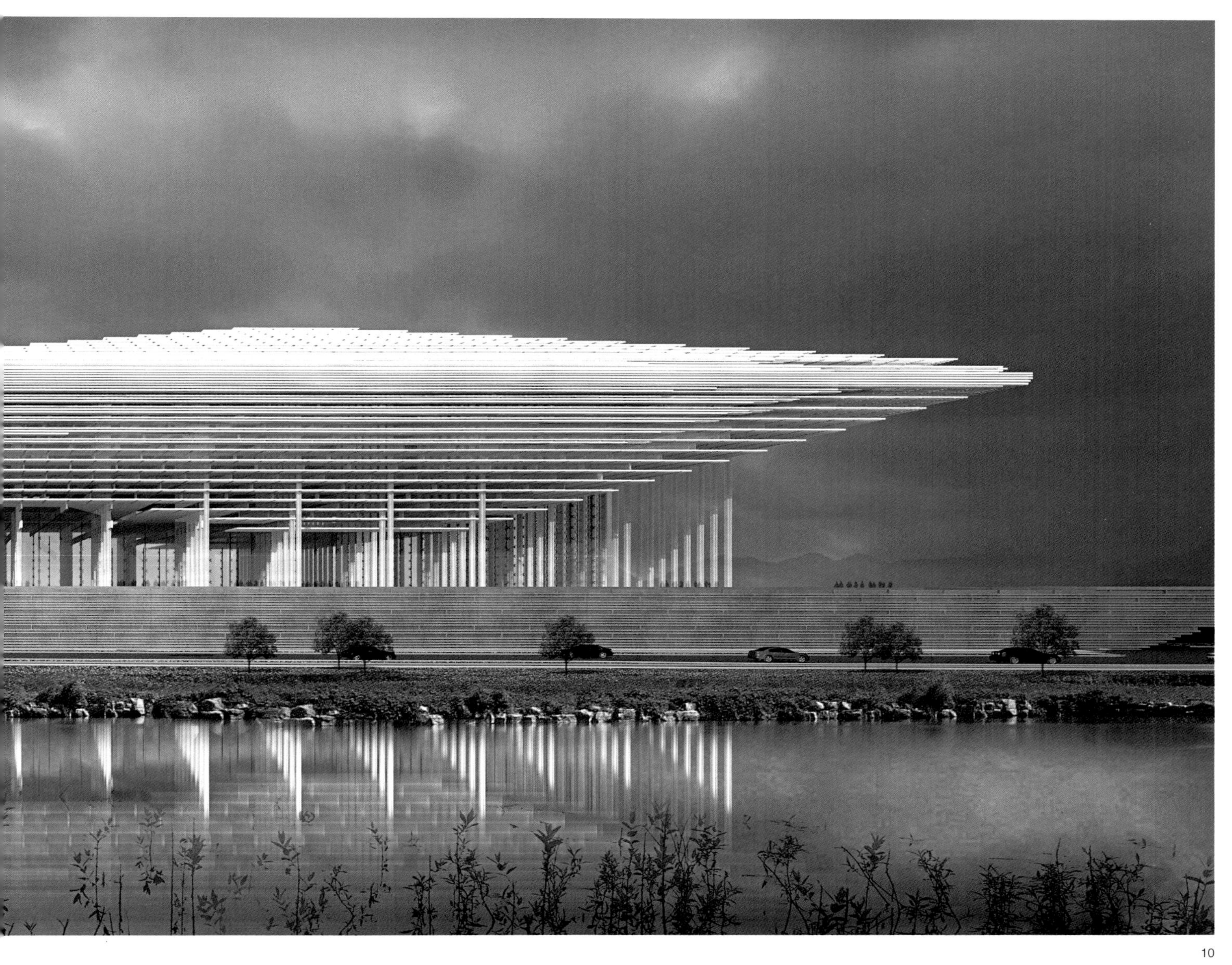

10

10 建筑西立面
West elevation

中国科学院国家科学图书馆 北京

National Science Library of CAS Beijing

设计时间 / Design：1999
建成时间 / Completion：2002
建筑面积 / Building area：41000m²
项目组 / Design team：崔彤 白小菁 夏炜 王知非 宋爽 邝红军
业主 / Client：中科院文献情报中心
摄影 / Photographer：杨超英 傅兴

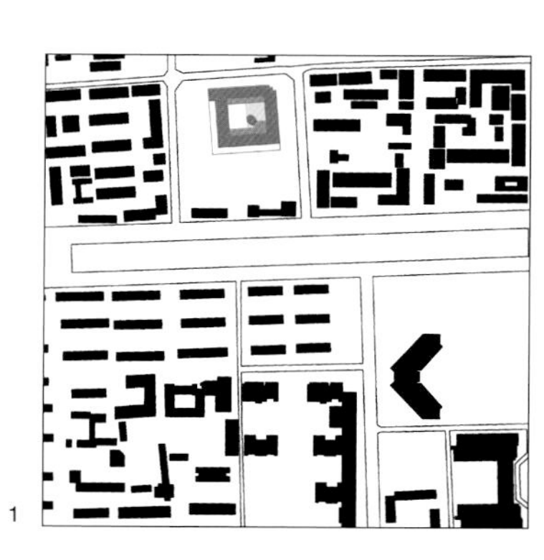

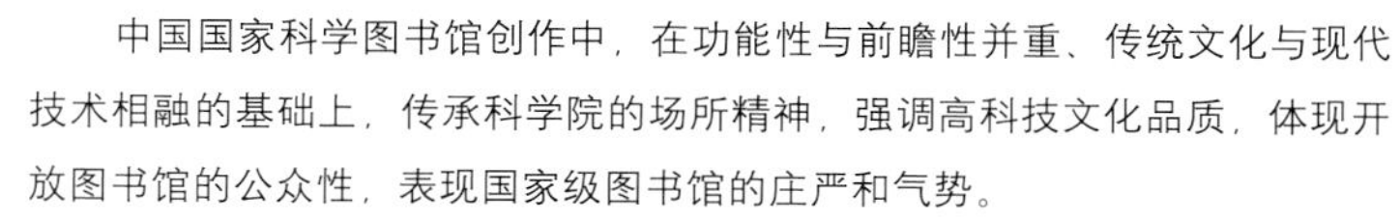

中国国家科学图书馆创作中，在功能性与前瞻性并重、传统文化与现代技术相融的基础上，传承科学院的场所精神，强调高科技文化品质，体现开放图书馆的公众性，表现国家级图书馆的庄严和气势。

建筑整体围绕着一个向西南开放的内院展开，内院的逻辑源于对传统四合院背后意义的新解释，它带给建筑能量：阳光、空气、景观，这便是方案所追求的新模式：集中式的便捷，分散式的环境优雅，被融汇在一个理性的平面中，从而确保阅览空间的采光通风。

进入图书馆的序列被有意设计成阅览建筑的过程，在此复辟了中国传统空间的叙述性，使"走建筑"、"读建筑"成为可能，一系列重要事件的连续发生犹如传统空间序列，其中内院正对的庙宇已成为知识的殿堂，而这一过程并没有神的意味，更多的是开放的图书馆显示的文化品质。

The design for the National Science Library of China inherits the sense of place from the Chinese Academy of Sciences, emphasizing high-tech cultural quality, reflecting the public characteristics of an open library, and showing the nobility and momentum of the state level library on the basis of paying equal attention to function and perspective, as well as a harmonious combination of traditional culture and modern technology.

The whole building surrounds an inner courtyard from south to west, giving a new interpretation of the traditional siheyuan (a courtyard dwelling). It brings energy to the building, such as sunshine, air and landscape, forming a new design that attempts to pursue convenience in a centralized way, and a graceful environment in a discreet way. All these are incorporated in a rational plane to ensure sufficient lighting and ventilation to the reading space.

The sequence of entering the library is intentionally designed to be a process of viewing the building's architecture, where it restores the narration of Chinese traditional space and makes it possible to "read the building when going through it." A series of important events seems to be the traditional space sequence, of which the temple against the inner courtyard to the right becomes a place of learning. However, such a process does not contain any religious meaning; instead, it presents the more cultural qualities of the library as an open and public venue.

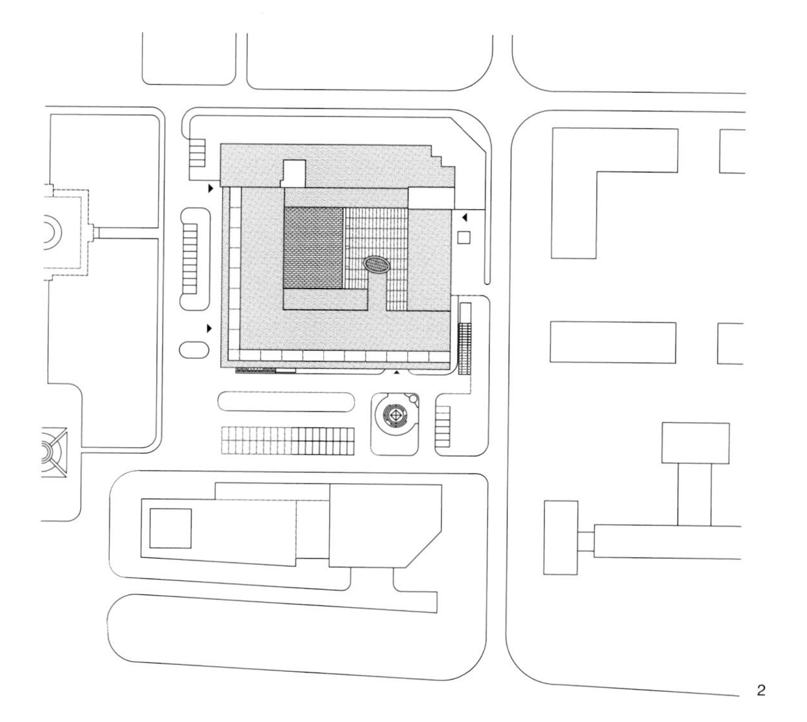

1 区位图
Location

2 总平面图
Site plan

3 建筑外景 1
Exterior 1

3

4

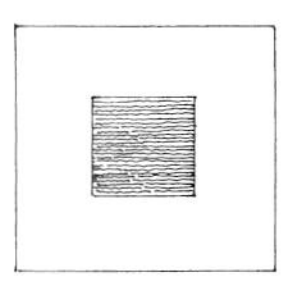
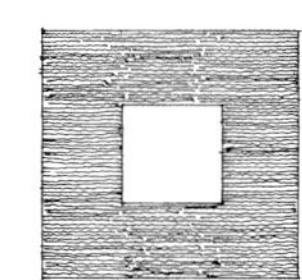
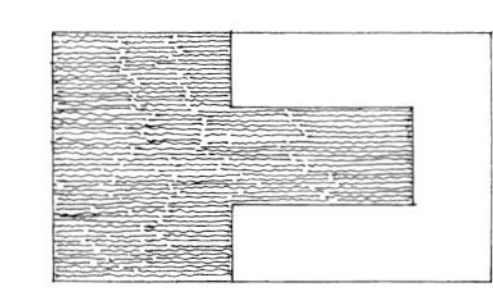

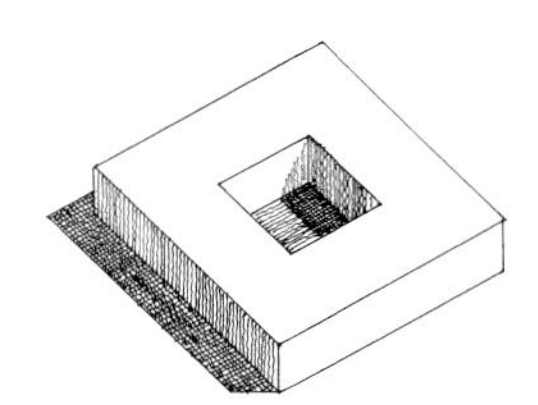
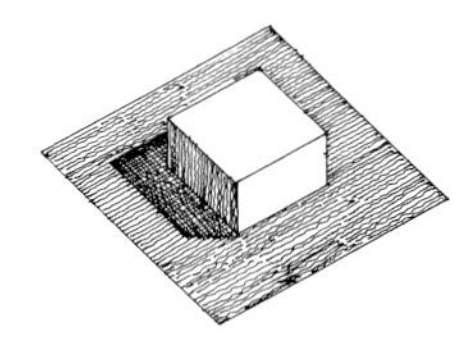
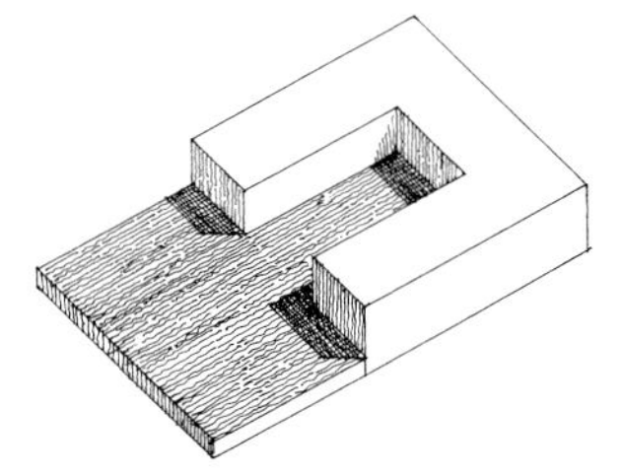

a 建筑包着花园
Garden in the building

b 花园包着建筑
Building in the garden

c 建筑嵌入环境
Building embedded in the garden

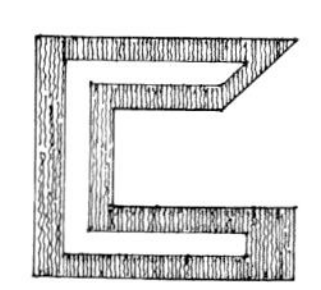
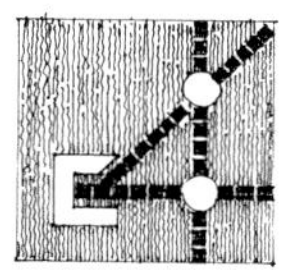
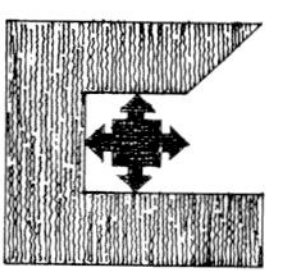

d 三个基本体块
Three basic blocks

e 两条主要轴线
Two main axises

f 一个中心庭院
One courtyard

5

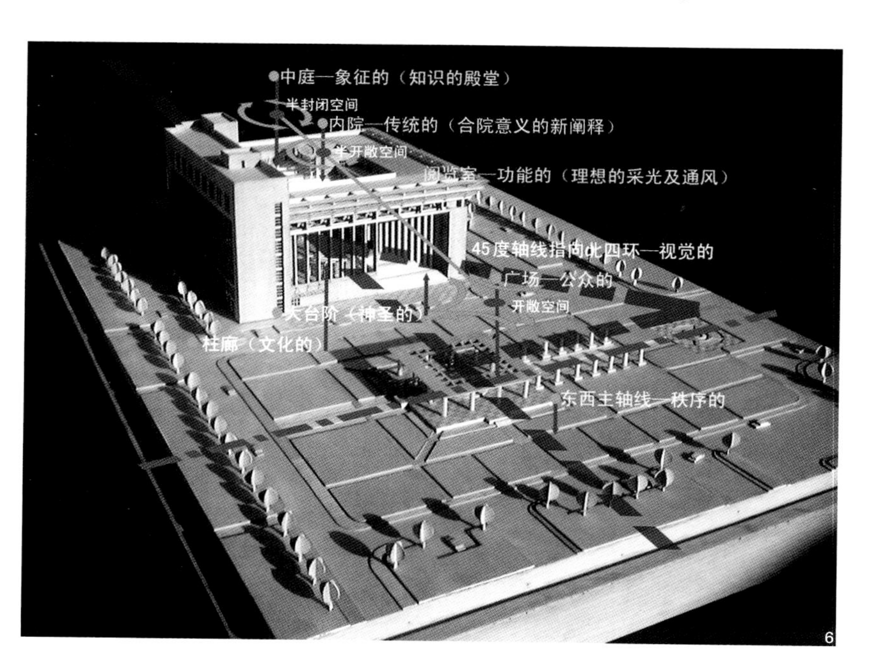

6

4 设计手稿 1
Hand Sketch

5 设计构思
Concept

6 模型
Model

7 建筑外景 2
Exterior 2

7

8

8　建筑外景 3
Exterior 3

9

9 建筑外景 4
Exterior 4

10 首层平面图
First floor

11 二层平面图
Second floor

12 三 / 四层平面图
Third/fourth foor

13 五层平面图
Fifth floor

14 六层平面图
Sixth floor

15 七层平面图
Seventh foor

16 剖面 1-1
Section 1-1

17 剖面 2-2
Section 2-2

18 剖面 3-3
Section 3-3

19 南立面图
South elevation

20 北立面图
North elevation

21 东立面图
East elevation

22 建筑外景 5
Exterior 5

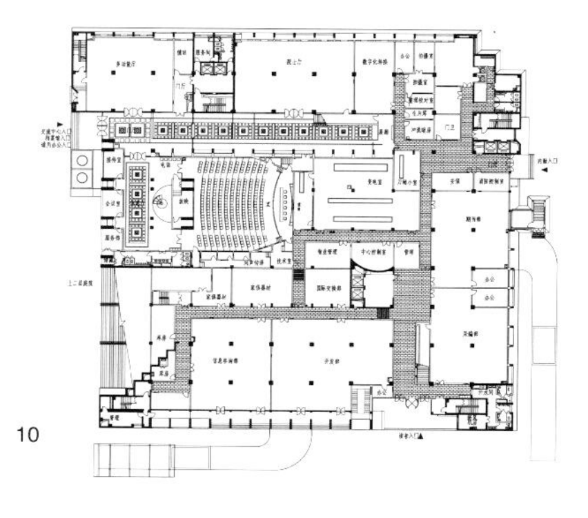

10

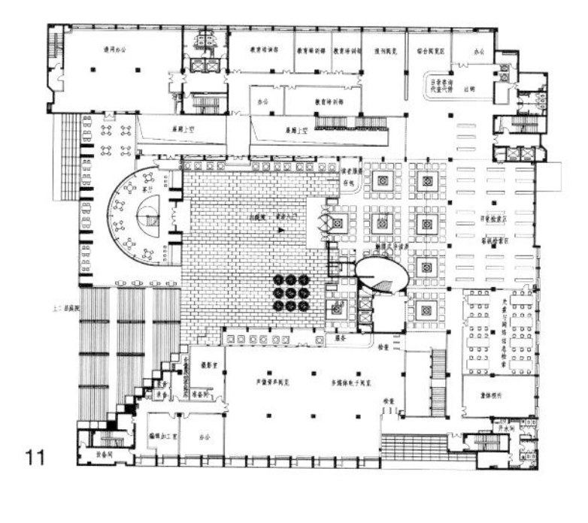

11

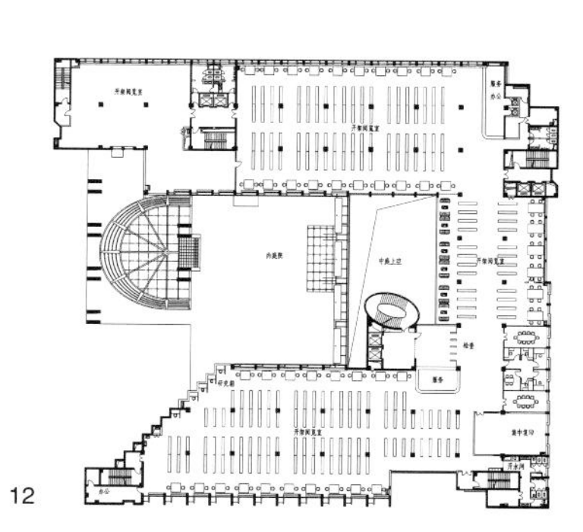

12

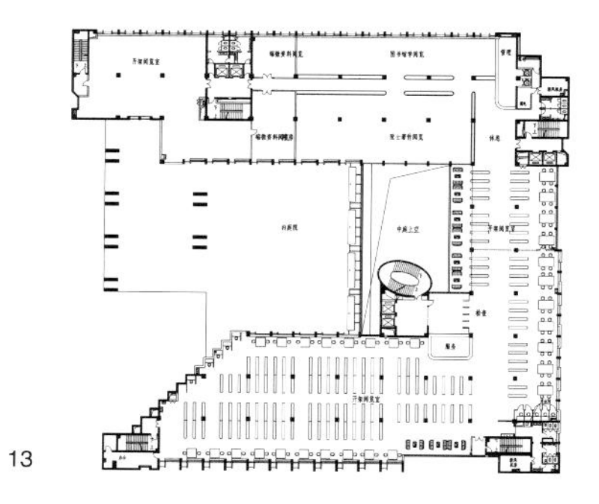

13

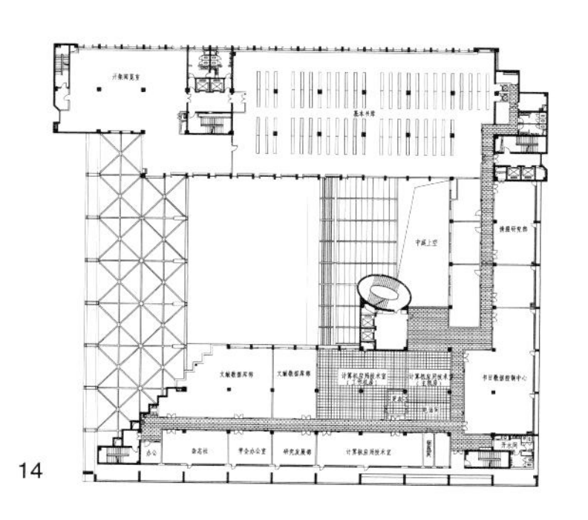

14

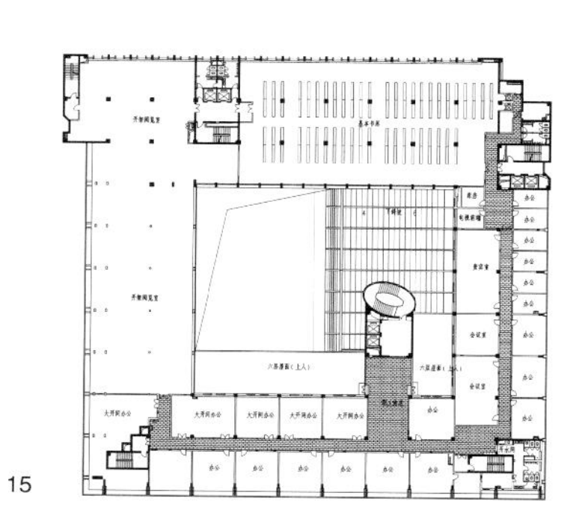

15

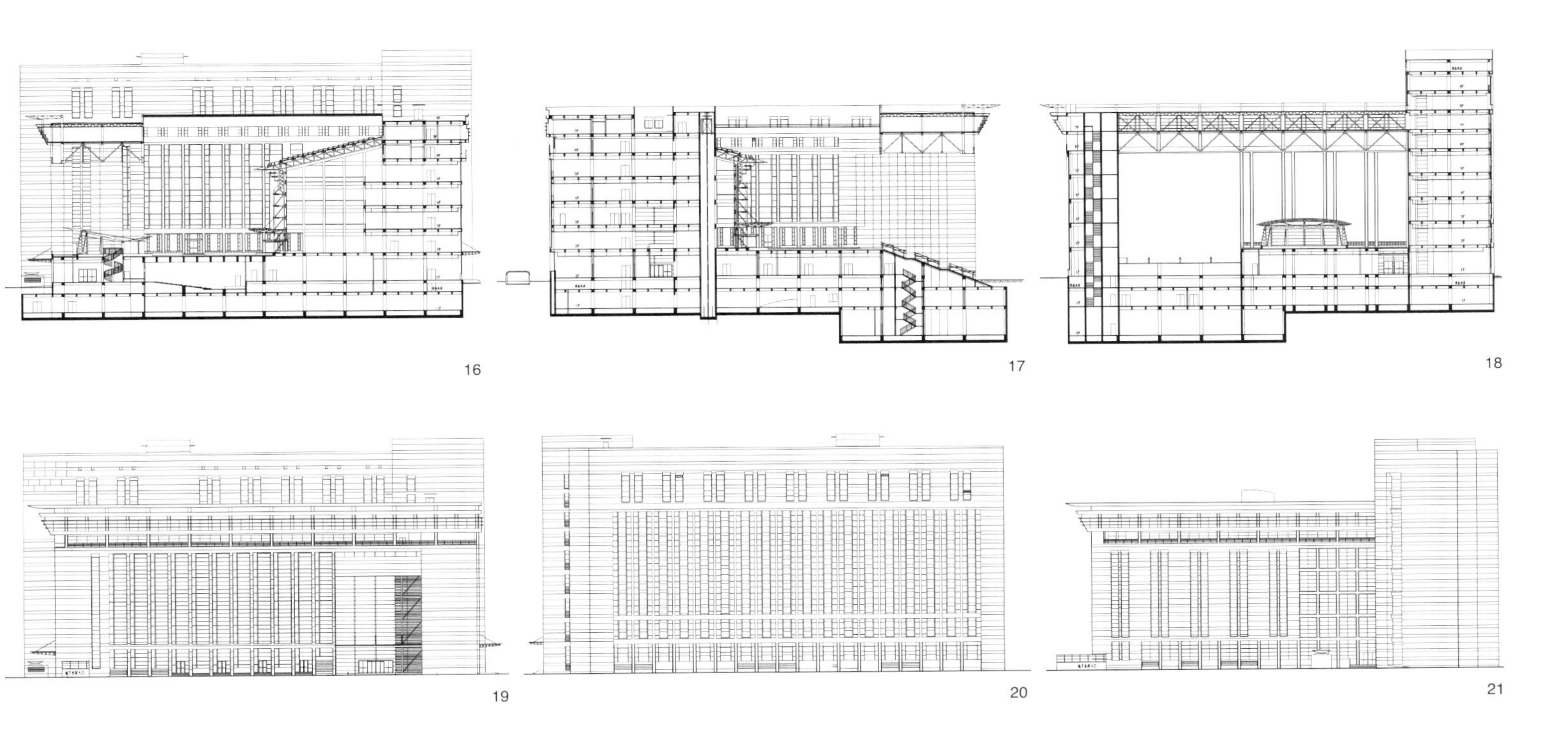
16 17 18
19 20 21

22

23

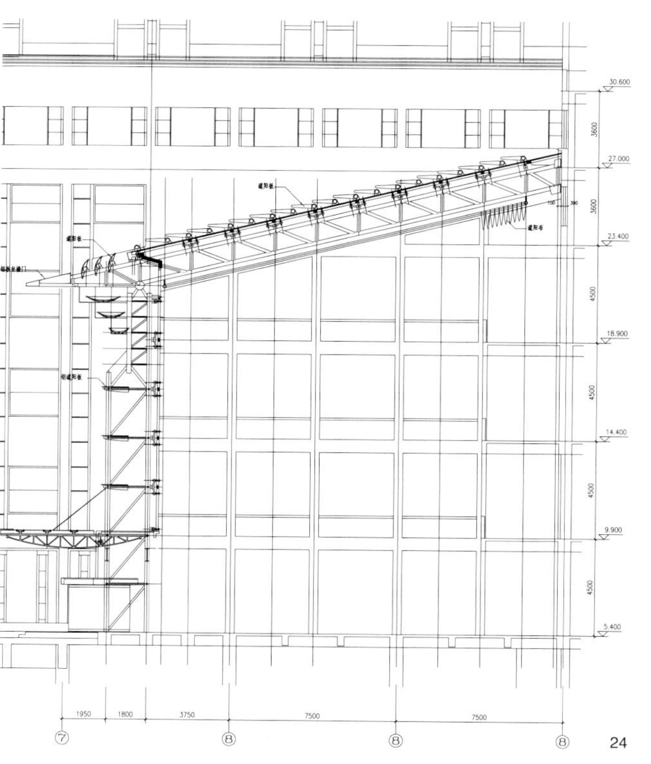

24

23 建筑外景 6
Exterior 6

24 幕墙节点
Detail

25 建筑外景 7
Exterior 7

25

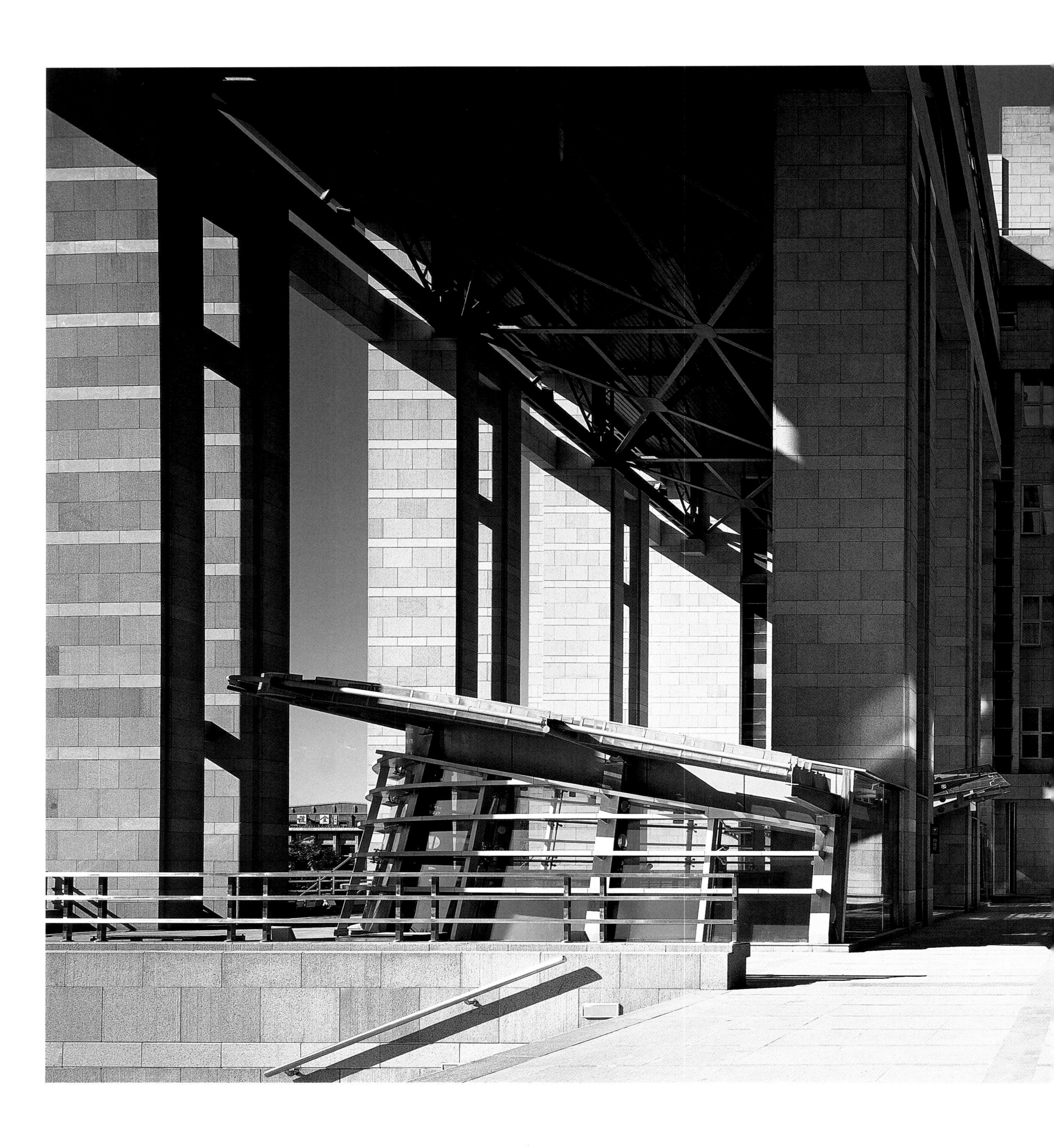

26

26 建筑外景 8
Exterior 8

北京林业大学学研中心 北京

Academic Research Center of Beijing Forestry University Beijing

设计时间 / Design：2008
建成时间 / Completion：2013
建筑面积 / Building area：96000m²
项目组 / Design team：崔彤 何川 罗大坤 苏东坡 张润欣 华夫荣 桂喆 唐璐 池依娜 辛鑫
业主 / Client：北京林业大学
摄影 / Photographer：杨超英 范虹

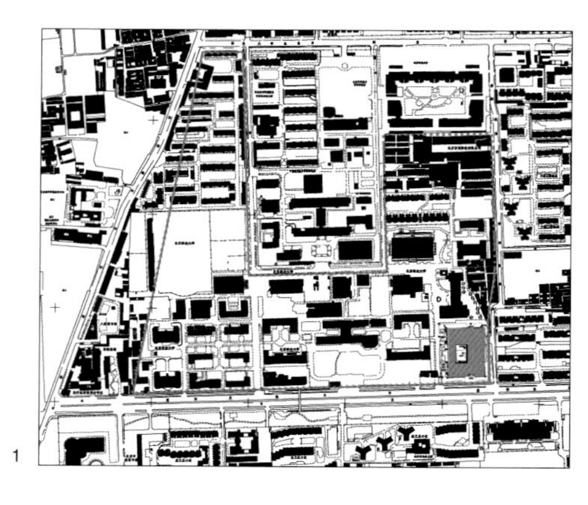

1 区位图 Location

2 建筑外景 1 Exterior 1

学研中心位于林业大学校园的东南角，它的双重性不仅表现在面对清华东路的城市表情和回应校园的亲和力，而更为重要的是它在“角部”的控制力以及重构校园秩序的新策略。

“U”形建筑以嵌入式的外部空间亲和于校园，构成一个静谧的“人文书院”。布局中南翼为院系综合办公，北翼为教学实验楼，东翼是阶梯教室及研讨教室，顶层布置了高端学子研讨及展览功能，地下一层作为一个特殊的功能单元，包括图书馆报告厅、展览等内容。

“U”形建筑限定了一个空间范围，构成内向性场所，同时朝外指向西侧校园。“U”形建筑在实现自身合理性同时成为林大校园东南角的收束而具有终极感，因此连接校园的东西轴线显得格外重要，西向敏感性相对于其他三个面具有独特的地位，这不仅在于它允许该范围与相邻的空间保持视觉上和空间上的连续性，而且在于人流动线由西而东，因此，西向界面在保持完整同时尽量划分适宜以取得与校园尺度的一致性。

“书院”作为轴线尽端的节点，是一个“空”的中心，正是由于教学空间之间的“空”所具有“弹性”和“聚合力”使得它成为区别于内与外的精神场所，在这样一个没有屋顶、三边围合、一边限定的方体空间中，东向的形态成为这个终极的要点，对于校园而言它是一个有力的底景；“书院”的多重性最终被凝聚在升腾而起的“树塔”之中，空间的多层次变化和纵深感给这个有限的“空”以无限的想象。

作为建筑的核心“方体内院空间”不仅要体现出对知识的神往；更多阐释了人与自然的共生理念；自然作为能量源给予内院空间风、光、热的同时也铸就一个可以释放出自然狂野能量的建筑；一个依附于土地可以汲取营养的、一个新陈代谢可以茁壮成长的、一个不断地衍生可以进化的“空间”，在建筑果壳内孕育着一种生机在奋进的成长中构筑成栋梁。一个“发芽种子”演化为“树塔”最终成为大学的精神图腾！

建筑风格选择了一条“中性”路线，建筑色调与主楼趋同，对外保持校园沿街的统一性；对内调和着红砖和灰砖的老建筑。均质化的竖向开窗和暖灰色的石材温和儒雅隐约显现着经典校园气质。对外简约卓尔不凡，对内谦和包容。“十年树木，百年树人”是重构场所经神的基础，建筑形态在延续校园空间结构中，吸取了一种繁衍的力量并建构了一个新的秩序化空间；纵横两条轴线控制了方正的体量，从两条轴线生长出沿街和内庭院的主入口，通过对自然律动的表达，借用分形几何学的手法创造出一种独特的形式语言。

The Academy Research Center is located at the southeast corner of Beijing Forestry University. Its dual nature is reflected on the city by facing Qinghua East Road and having an affinity with the campus, but even more important is its controlling force over its ″corner″ and a new strategy of reconstructing campus order.

The ″U″ -shaped building with its inserted external space is in accordance with the campus constituting a quiet ″humane college″ . The south wing of the layout is a general office building of departments, the north wing is a teaching and lab building, and the east wing is a lecture theater and seminar classrooms. The top floor features rooms for high-end studies and exhibitions, and the basement is a special functional unit including the library, auditorium and exhibition space.

The ″U″ -shaped building restricts spatial dimension by forming an internalized location while pointing to the campus on the western side. The ″U″ -shaped building is ultimately realizing self-rationality and brings a close to the southeast corner of the campus. Therefore the east-west axis connecting the campus is very important. Western sensibility is unique relative to the other three directions, not only because it allows a visual and spatial continuity with neighboring spaces, but also because the pedestrian flow is from west to east. For this reason, the western interface shall keep its integrity while being reasonable divided to be in harmony with the campus dimensions.

As the node at the end of the axis, the ″college″ has a ″vacant″ center. The ″vacancy″ with flexibility and aggregation between spaces makes it different from internal and external spirit spaces. In this square space without a three-side enclosure and one-side limitation, the eastward pattern is the key point of the ultimatum, being a vigorous rendering view for the campus; multiplicity of the ″college″ is finally amassed into the rising ″tree tower″ , which has multi-levels and depth and gives this limited ″vacancy″ limitless imagination.

2

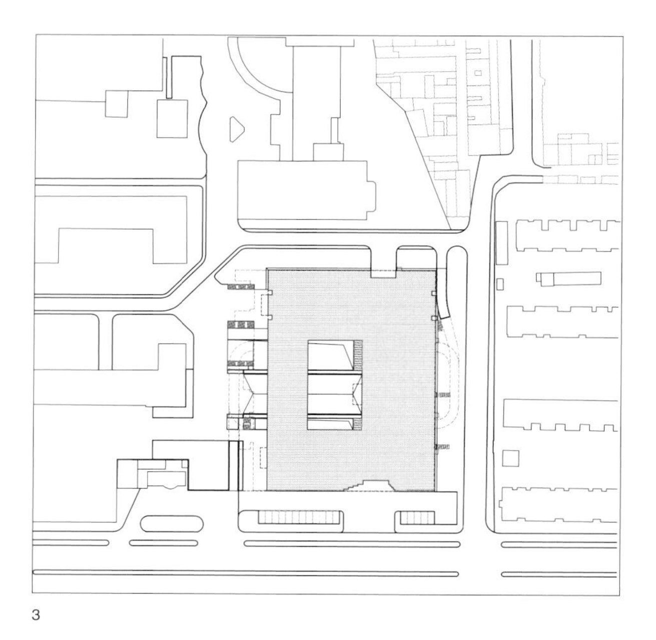

3

As the core of the building, the "square internal space" shall not only reflect the yearning for knowledge, but also explain the interdependence of human and nature; nature supplies wind, light and heat for the internal space as well as casts a building releasing natural brutal energy; the space which extracts nourishment from the land, grows up strongly and sturdily with its extension and can continuously develop to cultivate vitality within the building and becomes the core. A "grown bud" grows up to be a "tree tower" and finally becomes a spiritual totem of Beijing Forestry University.

The architectural style is chosen with neutral features and the building color is in accordance with the main building to keep the uniformity of the campus along the road; internally, it's an old building with red and gray bricks. Homogenized vertical windows and warm gray stone express a classical campus quality. It is simplified and exceptional seen from outside and it seems meek and contained. The Chinese saying that "it takes 10 years to grow trees, but 100 years to nurture a generation of good men" is the spiritual basis of the reconstruction. The architectural form absorbs conception from the continuous campus structure to construct a new ordered space; the vertical and horizontal axes control the upright size. The two axes extend along the road and an entrance to the internal yard. It creates a unique formal language by expressing natural rhythms and using principles of fractal geometry.

4

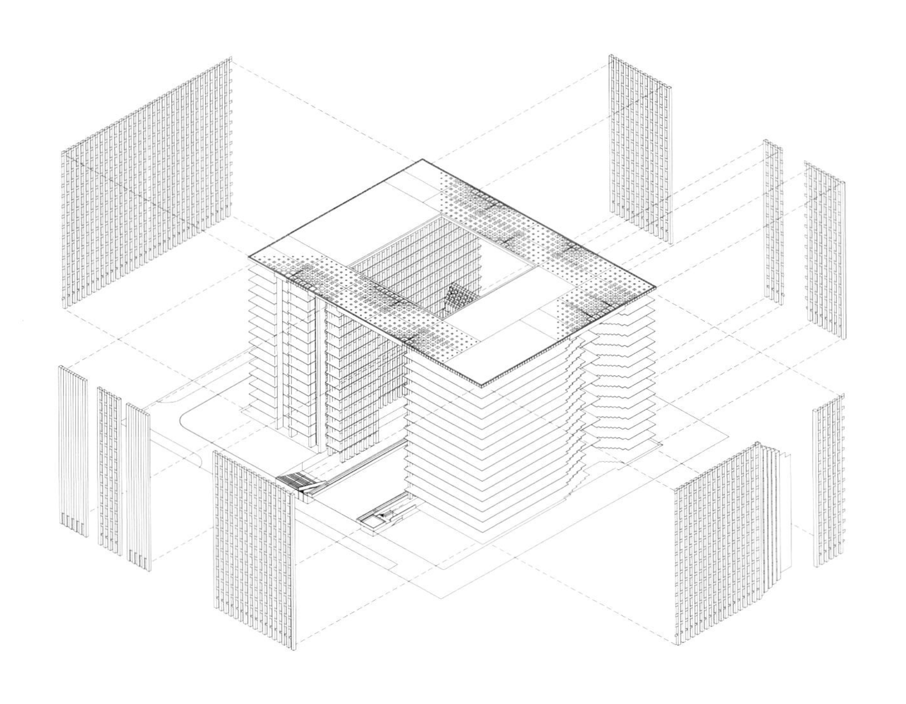

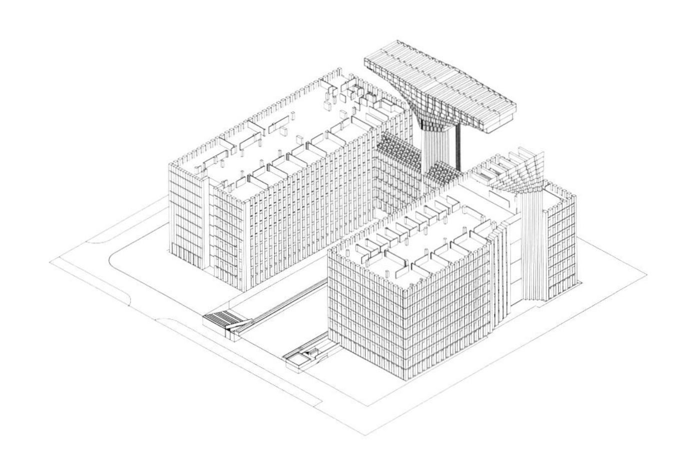

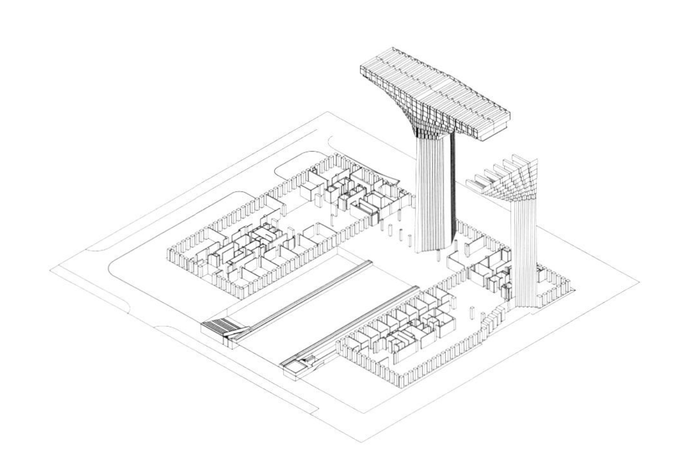

5

3 总平面图
Site plan

4 建筑外景 2
Exterior 2

5 解析模型
Analytic model

6　建筑外景 3
Exterior 3

6

7

8

9

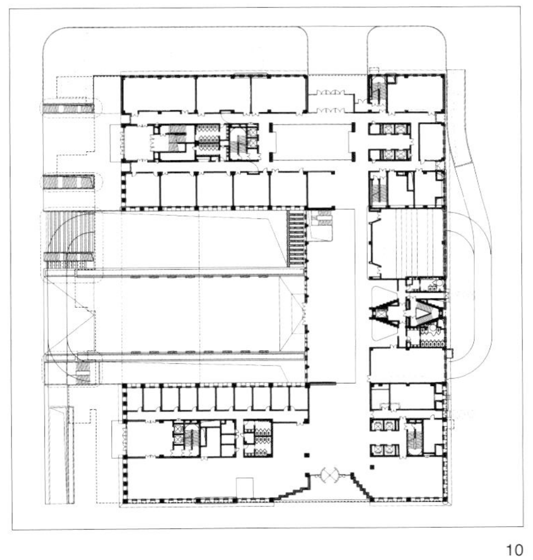

10

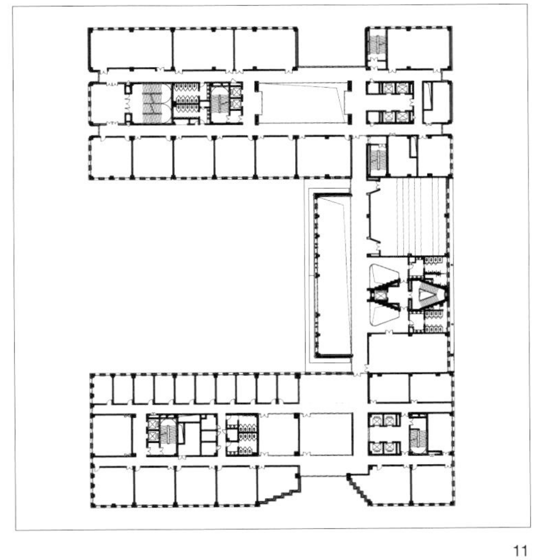

11

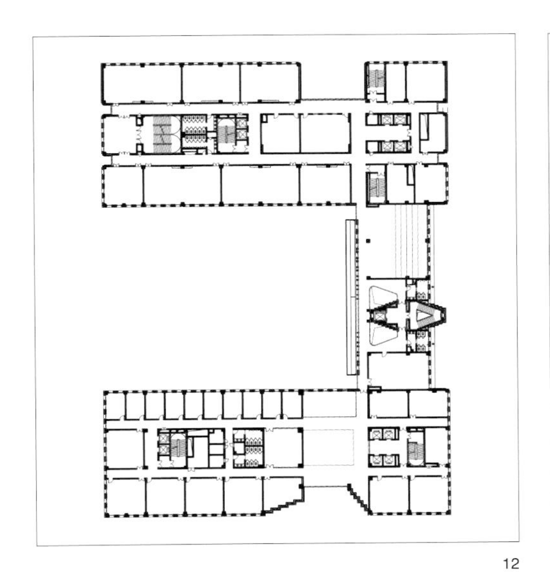

12

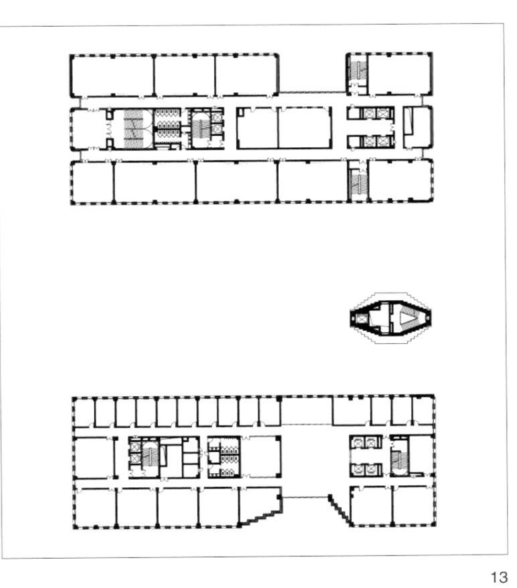

13

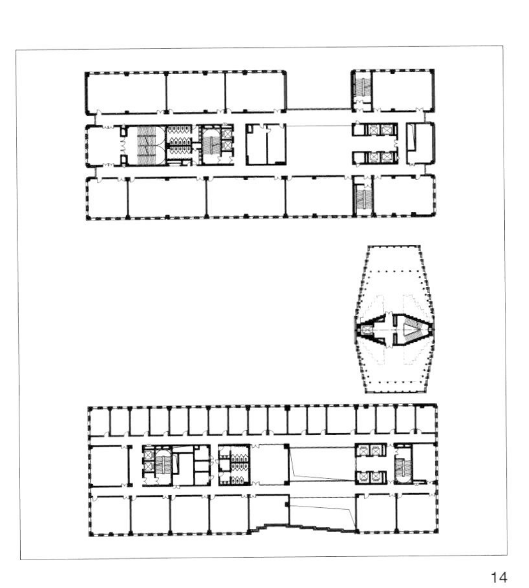

14

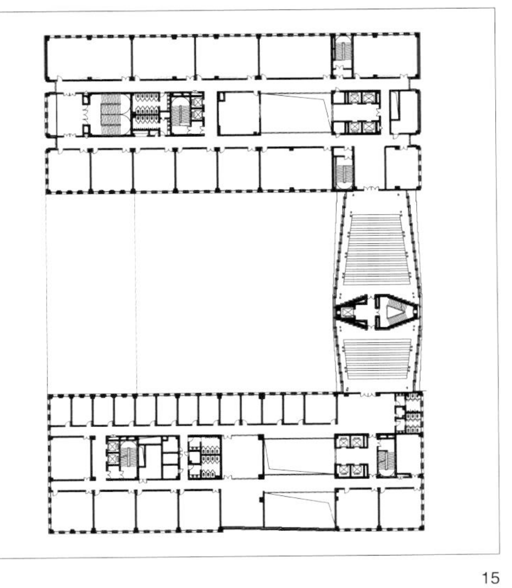

15

7 建筑外景 4
Exterior 4

8 建筑外景 5
Exterior 5

9 建筑内景 1
Interior 1

10 首层平面图
The 1st floor plan

11 三层平面图
The 3rd floor plan

12 七层平面图
The 7th floor plan

13 十层平面图
The 10th floor plan

14 十三层平面图
The 13th floor plan

15 十四层平面图
The 14th floor plan

16

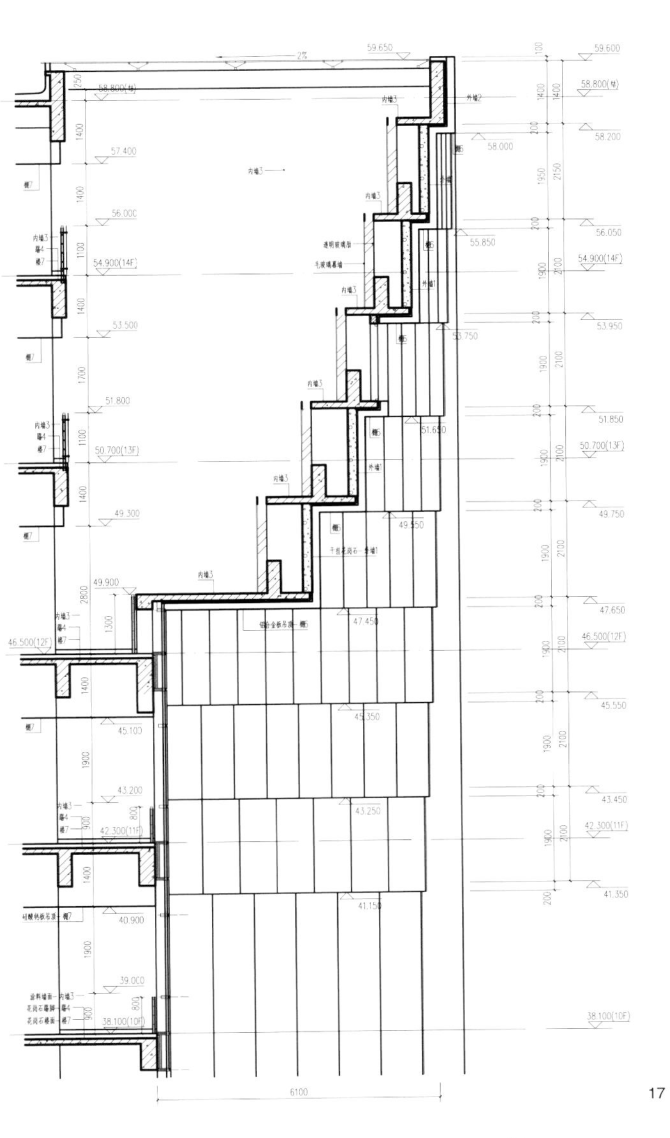

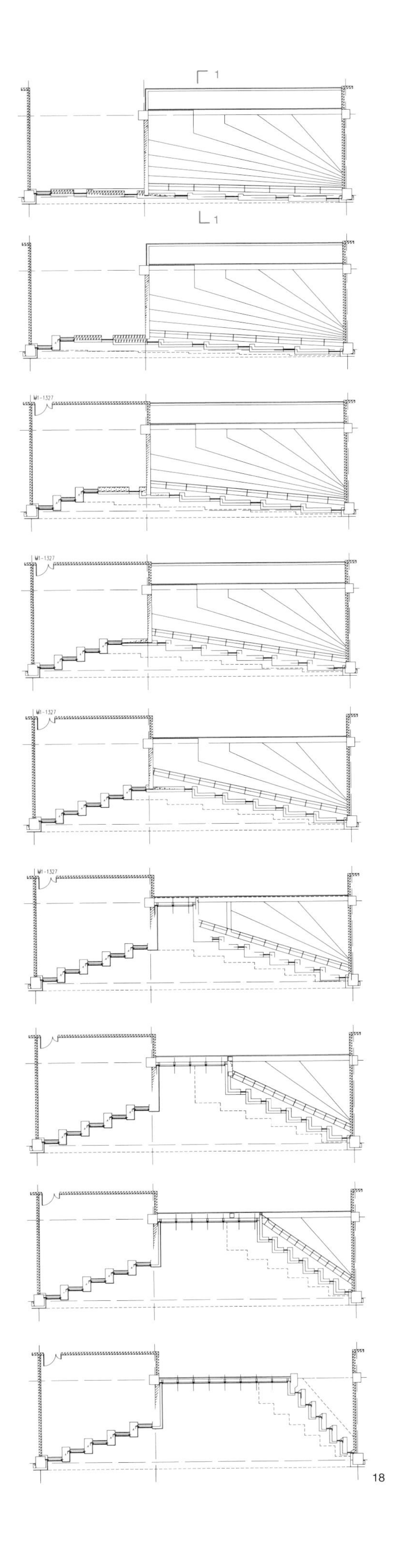

16 建筑外景 6
Exterior 6

17 剖面 1-1
section 1-1

18 节点 1
Detail 1

19 节点 2
Detail 2

20 剖面 2-2
Section 2-2

21 剖面 3-3
Section 3-3

22 剖面 4-4
Section 4-4

23 建筑外景 7
Exterior 7

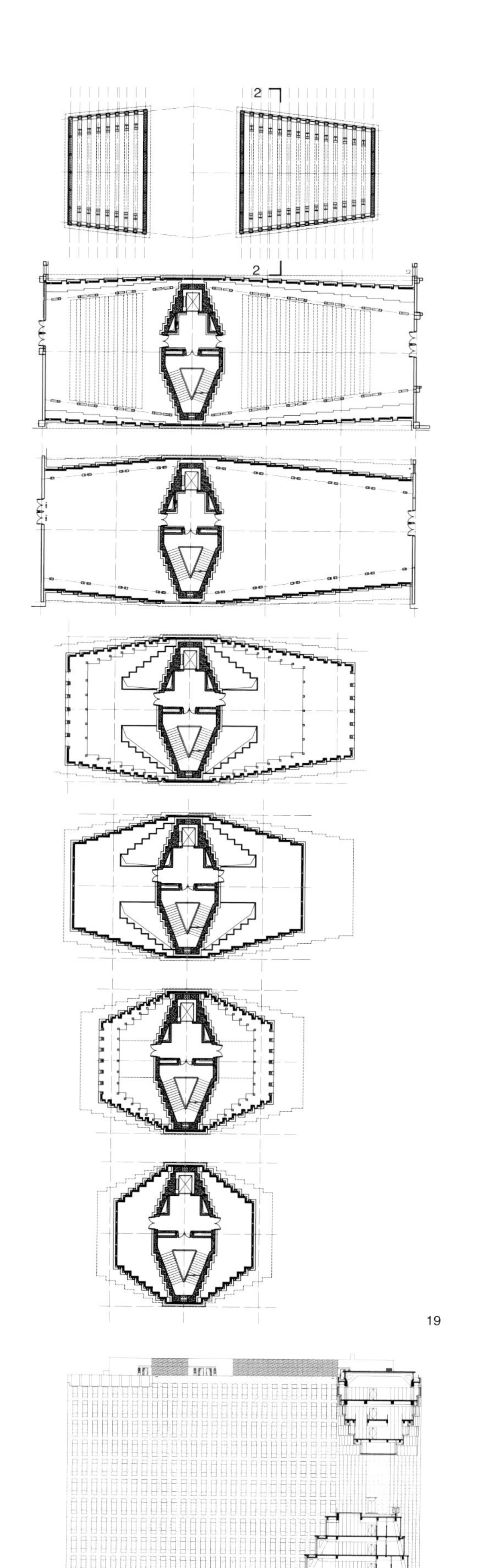

19

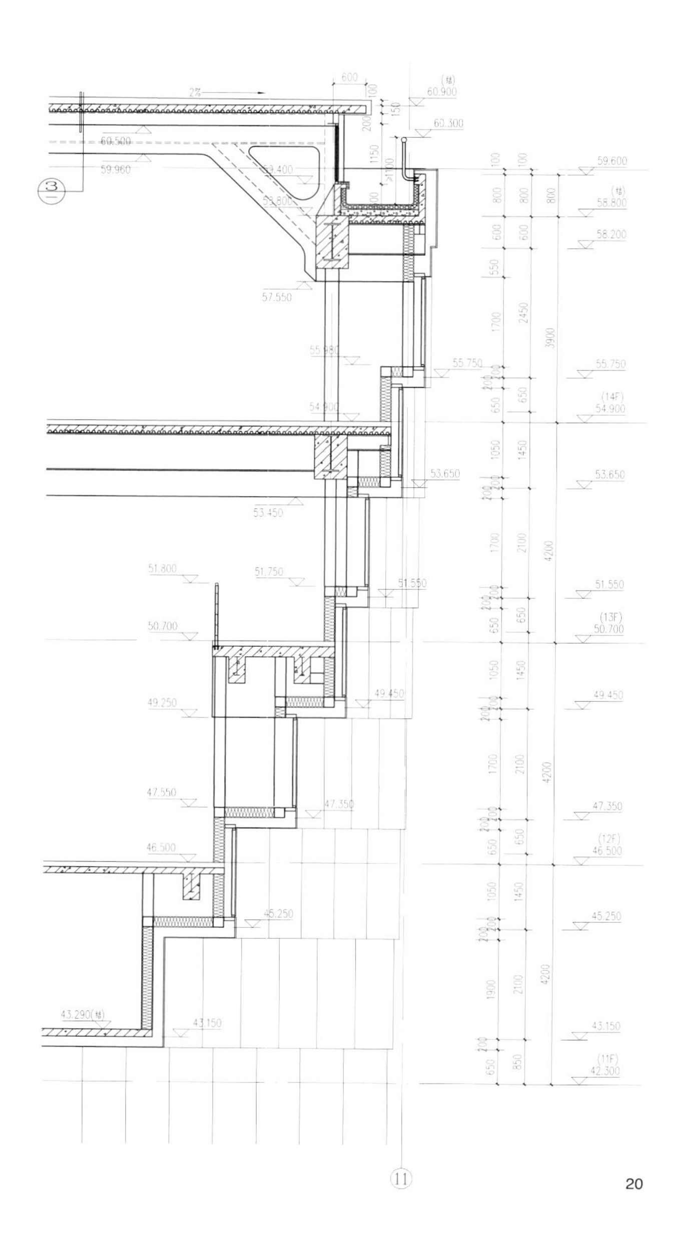

20

21

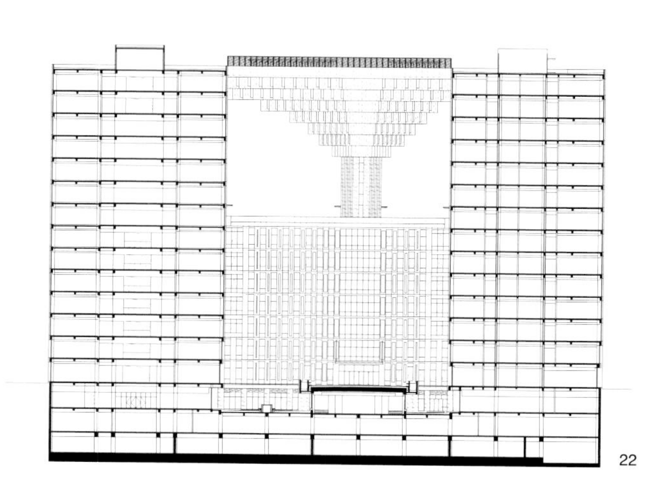

22

23

中国科学院化学所前沿交叉研究平台 B 楼 北京

Macromolecule Research Building B of Chemical Institute, CAS Beijing

1 区位图
Location

2 建筑外景 1
Exterior 1

设计时间 / Design：2008
建成时间 / Completion：2012
建筑面积 / Building area：23400m²
项目组 / Design team：崔彤 何川 张润欣 刘立森 黄文龙 苏东坡 华夫荣 刘建平
业主 / Client：中国科学院化学研究所
摄影 / Photographer：杨超英

项目位于北京市海淀区中关村北一街 2 号，中关村五号园区内，毗邻北大、清华等著名学府以及中科院各科研院所，具有先天的文化与科学氛围。在设计中，打通并适当拓展区内原有主干道路，将用地自然划分为若干功能区域。各区域内车行、步行道路自成系统，同时又便于相互联系。

建筑平面为南北分区的〝C〞型布局，南侧为学术办公模块，北侧为实验研究模块，之间为服务模块，分区明晰，功能紧凑。办公模块设置小间研究室；实验模块主要为通用化学实验室；服务模块为建筑设备与生活卫生服务空间。报告厅位于一层内院，位置居中，服务便捷。

根据化学实验室自身的特点，采用了通用可展、灵活实用的模块化设计方案。管网系统集中布置，高效且易于检修，同时考虑先进的智控变频集中补排风系统，降低运行成本，节约能源。

建筑整体形象朴实无华，庄重文雅。红色面砖的外饰面与园区原有的建筑协调一致，作为主要立面语言的方形洞窗方便易用，同时也兼顾了园区内现有建筑的造型语汇，辅助砖百叶的使用，使园区建筑形象得到良好的整合。

建筑体量方正有力，呈现片状叠置效果，于西侧临街立面表现尤为明显，诠释了科研建筑严谨内敛的风貌，恰似一部部厚重的学术典籍，隐喻了科研机构求真、求知的历史使命。翻开〝学术典籍〞来到建筑东向内院，化学分子结构意向的装饰幕墙与入口网架营造出极富化学科研情趣的入口空间，使得化学科研的精神在此凝聚和提炼，最终实现了一个沉稳、静谧的研究空间。

The project is located at No.2 North First Street, Zhongguancun, Haidian District in Beijing. Located in No.5 Zhongguancun Science Park, it is adjacent to Peking University, Tsinghua University and other well-known universities and scientific research institutes of the Chinese Academy of Sciences, so it embodies a strong cultural and scientific atmosphere. In the design, the original main roads in the area are opened up and expanded to a certain extent, and the land is naturally divided into several functional areas. The roadways and walkways in each area are independent which can also be connected easily.

The building plan is the〝C〞-type layout with a north-south partition. The southern area is the academic office module, the northern area is the lab research module, and the area between is the service module. The clearly divided area has densely-arranged functions：the small research rooms are arranged in the office module; the lab module is mainly for the general chemical lab; the service module is planned for architectural equipment and life and health service spaces, and the report hall is in the interior yard of the ground floor, enabling convenient service.

Based on the features of the chemical lab, the module design scheme is adopted for general, extendable, flexible and practical use. The centralized pipeline network system is efficient and easy to maintain, and the advanced intelligent control and frequency-conversion centralized air supply and exhaust system is adopted to reduce operating costs and save energy.

The overall style of the building is simple and elegant. The exterior red brick finish is consistent with the existing buildings in the area, while the square openings and windows as the major façade language are convenient for use, and it also takes into consideration the vocabulary of the existing buildings in the area, supported by the brick shutter. The building style of the park appears to be integrated together.

The building is of a good size and powerful, showing an effect of sliced superposition. This is particularly true for the west façade facing the street, interpreting the style of precision and restraint of the scientific research building. Just like a massive academic tome, the building is a metaphor of the historical mission of truth-seeking and desire to learn. Unfolding the〝academic tome〞, one comes to the interior yard in the east direction of the building, and the decorative façade and the network frame at the entrance add the rich theme of chemical scientific research, of which spirit is condensed and refined here, to finally realize a stable and quite research space.

2

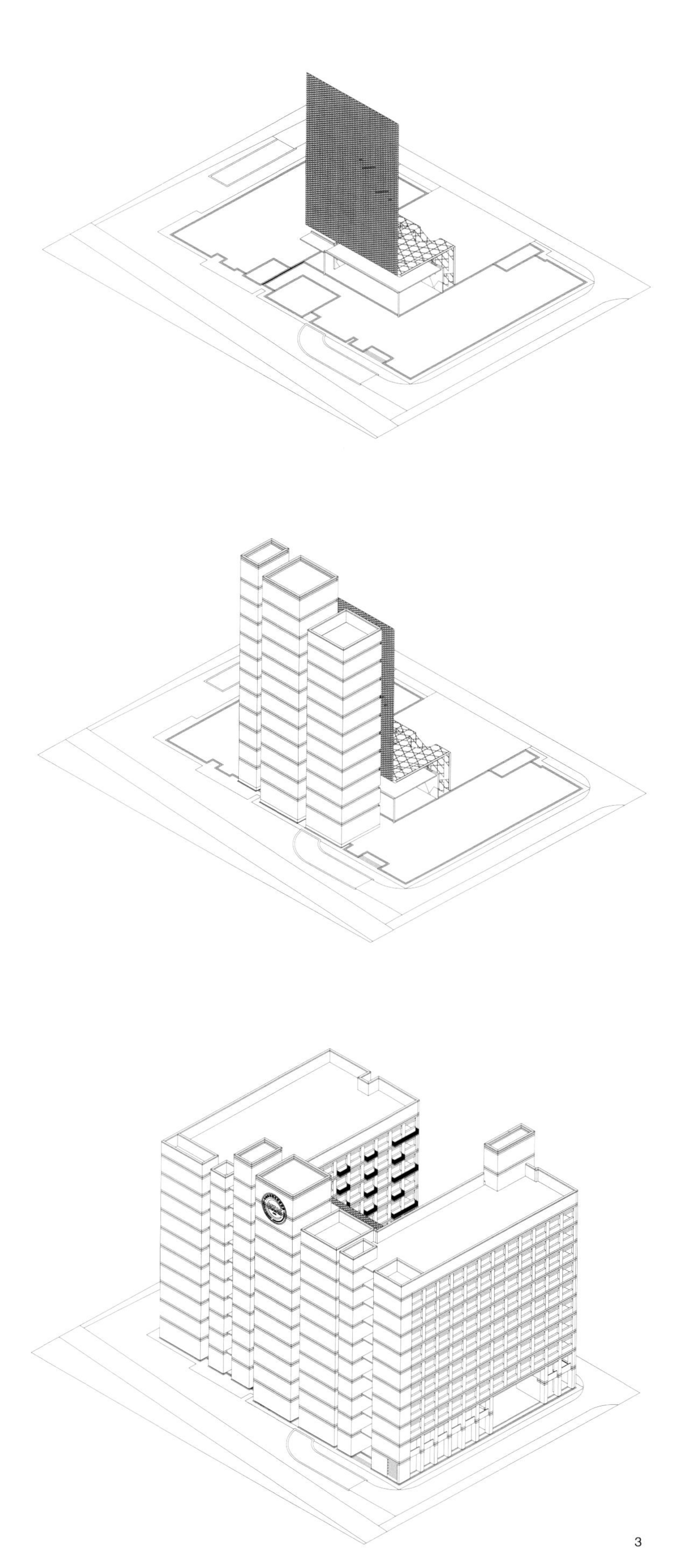

3

3 解析模型
Analytic model

4 建筑外景 2
Exterior 2

4

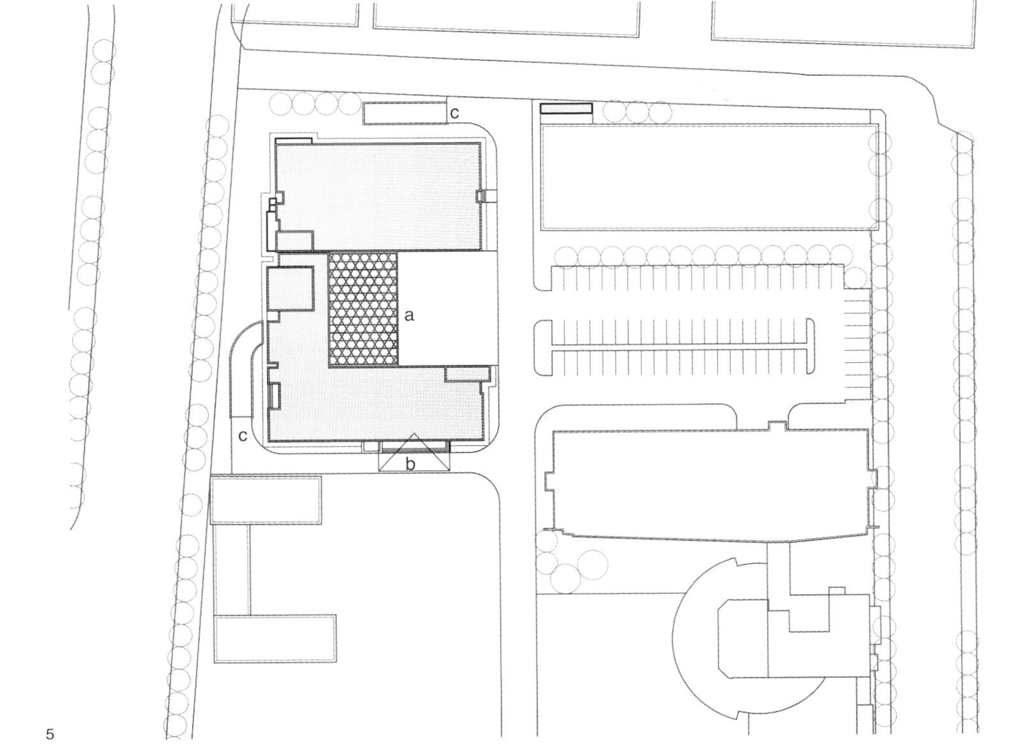

5

5 总平面图
Site plan

a 建筑主入口
Main entrance
b 建筑次入口
Auxiliary entrance
c 车库入口
Garage entrance

6

6 建筑外景 3
Exterior 3

7

7 建筑外景 4
Exterior 4

8 西立面图
West elevation

9 南立面图
South elevation

10 东立面图
East elevation

11 建筑外景 5
Exterior 5

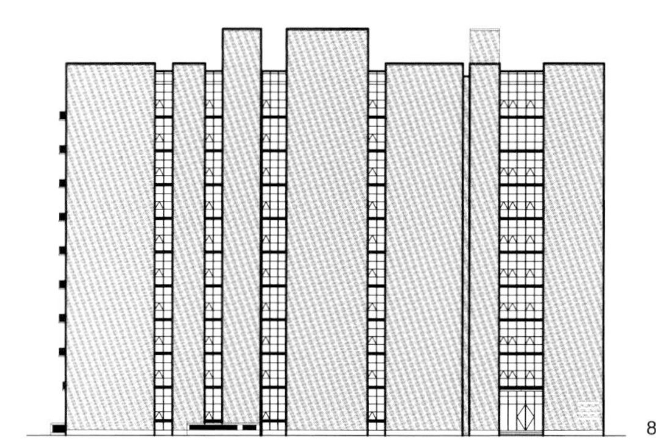

8

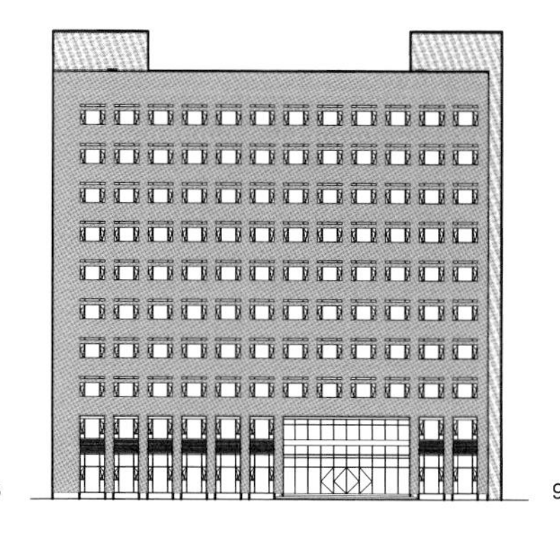

9

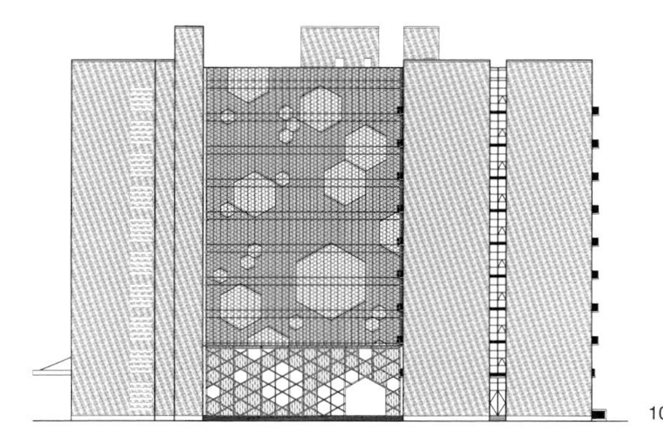

10

11

12

13

12 建筑外景 6
Exterior 6

13 建筑外景 7
Exterior 7

14 建筑外景 8
Exterior 8

15 首层平面图
The 1st floor plan

16 三层平面图
The 3rd floor plan

17 四层平面图
The 4th floor plan

14

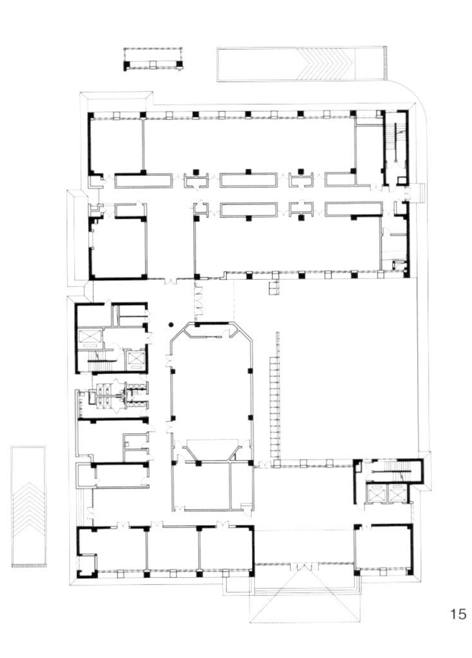

15

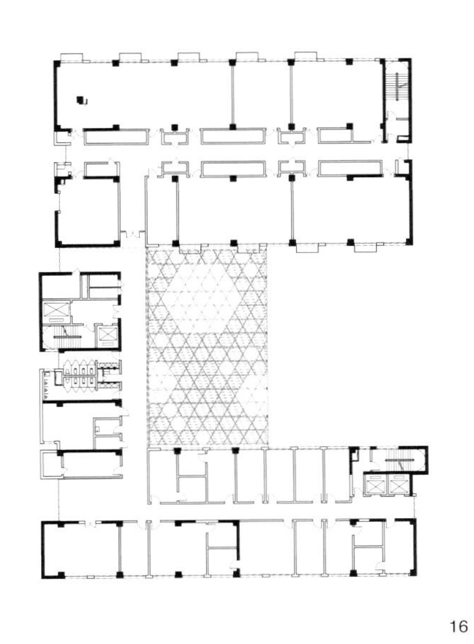

16

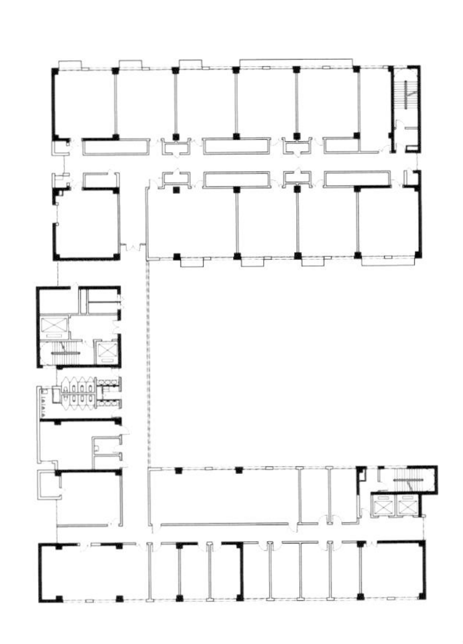

17

18

19

18 建筑细部 1
Exterior detail 1

19 建筑细部 2
Exterior detail 2

20 建筑细部 3
Exterior detail 3

20

中国科学院计算技术研究所 北京

Institute of Computing Technology, CAS Beijing

设计时间 / Design：2002
建成时间 / Completion：2004
建筑面积 / Building area：30000m²
项目组 / Design team：崔彤 王欣 李昕滨 文业清
业主 / Client：中国科学院计算所
摄影 / Photographer：杨超英 舒贺

1 区位图
Location

2 建筑外景 1
Exterior 1

本案位于联想园区，作为计算所先驱的中国科学院计算技术研究所在融于园区的同时，应该表现自身卓尔不凡的科技品质，以科研文化建筑为背景，在体现规划理性和秩序的同时，与中关村科学园区实现对话和整合，在更高更宽的视角中思考中科院计算所具有的独特品质。

结合计算所的使用功能，对计算机深层文化和建筑空间进行不懈的追求，形成功能性与前瞻性并重的独特科研办公建筑。设计的逻辑源于对计算机科研、教学、办公模式的功能解析，平面形态和空间秩序都是使用功能自内而外的反映。单元化、模块化的空间布局，源自使用功能的模数化设计，从1.4m×3 = 4.2m的研究员办公空间进深，到4.2m×2 = 8.4m的柱网，都是从工艺分析到技术实现的过程。南北两侧6.6m进深的研究员办公空间和实验室用房，中间15m的开敞计算机房和办公空间，分别对应于双走廊设计以及可灵活划分的科研教学单元。

为使得南北向采光通风资源的最大化，建筑的交通核心和设备用房布置在建筑的两端，匀质和对位整齐的核心形成高效便捷的交通体系网络，在确保内部可持续发展的同时，也为建筑设备的集中经济性布置创造了条件。

4层以上是内向型的科研模块，中间部分为效率型开放“工位”，南北两侧对应于科学家的研究单元，东西两端分别嵌入2层高生态仓，并构成独特创新场所。

从环境和工艺入手，坚持修正式的功能决定论，设计新模式回归了实用、经济、美观的最基本原则。设计求解的过程源于对计算所科研教学功能的新阐释：平面的模块化功能单元，纵向的逻辑功能分区以及从生物气候学的角度对建筑形态和材料的技术经济性比较，都使其成为可持续发展的建筑。设计目标是将单栋楼的设计转化为“立体园区”的创造，这就如同高度集成的芯片反映出的多功能复合化信息量的存储。设计方法是将不同的功能分层和切片，最终以物化的切片把功能块“焊接”成为高效率的“芯片”。

两层高的生态仓以人为本，设置一些绿色植物，用以吸收大量人群在建筑中呼吸释放的CO_2，提高室内的氧气含量，促进空气的流通，提供一个舒适的室内环境，利用建筑形体中部的跌落设计，形成半围合的屋顶花园。屋顶的绿色景观设计，为建筑提供了优良性能的结构保温层，既有利于建筑的保温隔热，也为使用者提供了方便宜人的休闲空间。

The Institute of Computing Technology of the Chinese Academy of Sciences is located at the Lenovo Science & Technology Park. As a leading institute built in the Park, its design should demonstrate its outstanding scientific quality. With the background of a scientific research culture building, not only should it reflect rationality and orderly planning, but should also realize communication and integration with the Zhongguancun Science Park, and reflect its unique characteristics on a higher and broader level.

Through the ceaseless pursuit of deeper cultural and architectural space of computers combined with the function of the computing institute, a unique scientific research and office building should be formed with equal focus on functionality and prospection. The logical design comes from the functional analysis of scientific computer research, teaching and office mode, and the plan form and spatial order are a reflection of the functional use from interior to exterior. The spatial functional layout based on "units" and "modules" comes from the module design of the function：the researcher's offices are 1.4m×3=4.2m and the column network is 4.2m×2=8.4m, which reflect a progress from process analysis to technical realization. The researcher's offices and laboratories have a depth of 6.6m in the north and south sides, with a 15-m open computer room and office space in the middle; they are intentionally designed to respond to the double corridor design and the scientific research and teaching units which can be divided flexibly.

The traffic center and plant rooms of the building are located on both sides of the building in order to maximize the lighting and ventilation resources. The uniform and symmetrical center forms an efficient and convenient traffic system network which can ensure internal sustainable development and create beneficial conditions for a centralized and economic arrangement of building equipment.

From the fourth floor and above, the building features an internal scientific research module; the middle part is an efficient open "work space"; the north and south sides serve as research units for scientists; and the east and west sides are embedded with a two-story high ecological store respectively, forming a unique and innovative place.

Starting from the environment and process and adhering to a determined modified functional philosophy, the new design mode returns to the most essential principle of practicality, economy and beauty. The solution process of design originates from the new interpretation of scientific research and teaching function of the institute of computing technology：the module

国科学院 算技术研究所

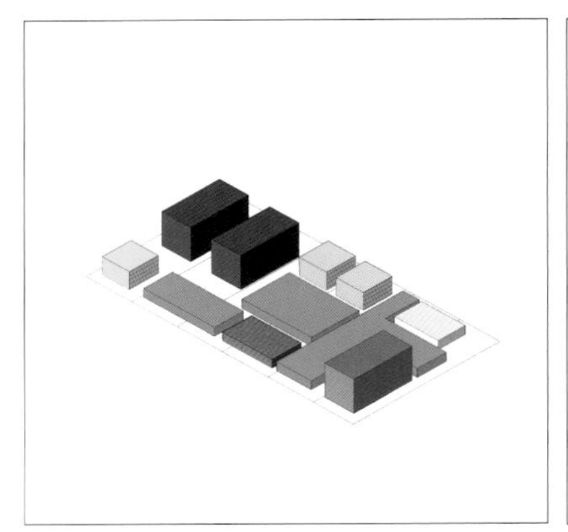
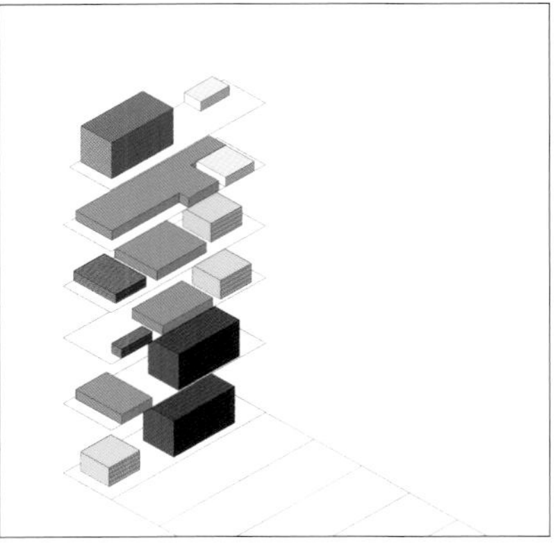
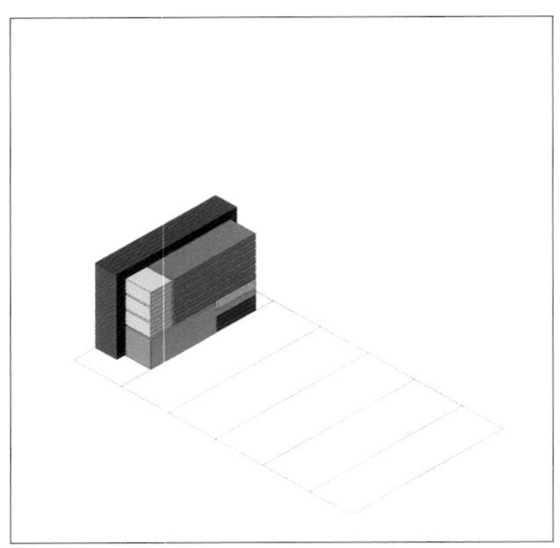
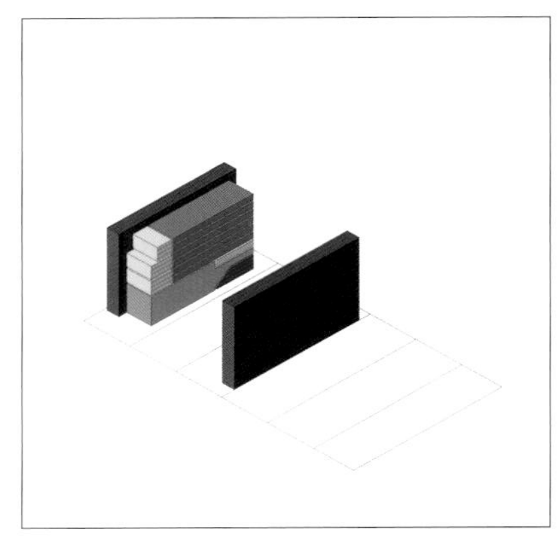
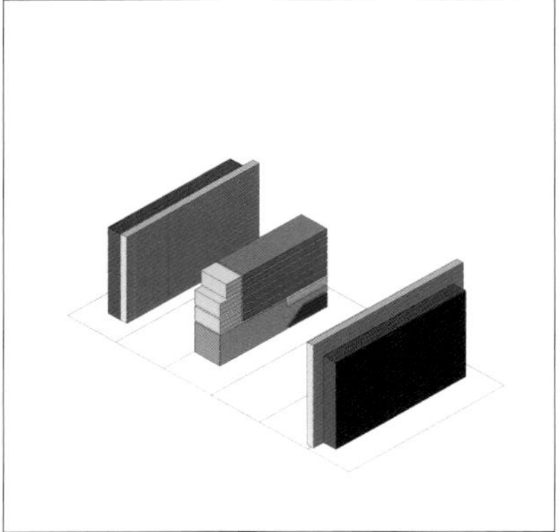
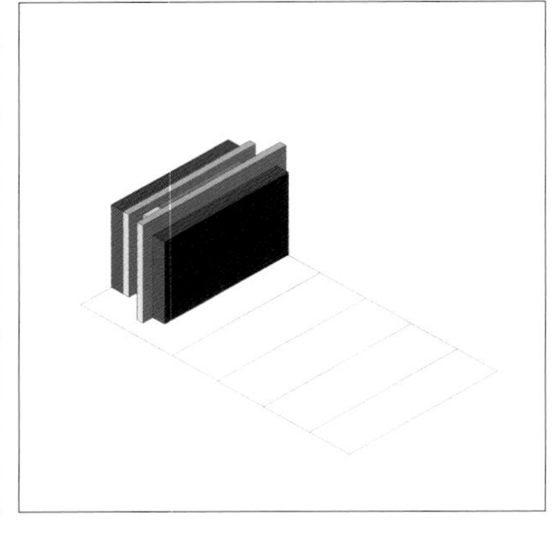

3

3 建筑解析
Analysis

4 建筑外景 2
Exterior 2

东西界面的设计在解析中国算盘和珠算技术的逻辑后，遵循自相似性原则形成跳动的数码信息图案，充分解蔽信息时代的网络技术和计算文化，而实现这种可能的建筑技术是隐框幕墙及条状丝网玻璃形成的细部特征和与众不同的影像建筑。

中国科学院计算技术研究所科研综合楼与联想园区建设环境的协调并非简单的吸收和移植，而是从场所中吸取片状的语言和东西轴的属性，继而转化为和而不同的建筑形态。平面强调南北布局的同时，强调东西轴线的延伸和景观视线的贯通，在中庭、大台阶、生态仓、屋顶花园和景观通廊，都可以远眺整个中关村科学园区，使建筑的内向性和开放性结合为统一的有机体，实现内心深处与周围环境的对话与交流。

functional unit based on the plan, vertical logical functional partition, as well as the technical and economic comparison of the architectural form and material from the perspective of nature; all these have made it a building of sustainable development. The goal of the design is to translate the individual building composition into the creation of a "three-dimensional park" , which is similar to that of a highly integrated chip reflecting multi-functional and complex information storage. The design method is to divide and cut different functions into different layers and slices, and finally, "weld" the functional block into an efficient "chip" with materialized slices.

After analyzing the logics of the Chinese abacus and its technology of calculation, the design of the east and west interfaces follows the similarity principle, forming a leaping digital information pattern, fully presenting the network technology and computing culture in the information era. The architectural technology to make this possible is the minute features formed by the hidden-framed curtain wall and strip-wired glass, creating a distinctive image of the building.

The coordination of the Scientific Research Building of the Institute of Computing Technology, Chinese Academy of Sciences with the environment of the Lenovo Science and Technology Park is not a case of simple absorption and transplantation. Instead, by absorbing the slice language and characteristics of the east-west axis from the site, it then is translated into an architectural style which is in harmony with the others. While stressing the south-north layout in the plan, it also emphasizes the extension of the east-west axis and the landscape has clean lines, so one can get a full view the entire Zhongguancun Science Park from the atrium, large steps, biological store, roof garden and landscape corridor. As a result, the interior feature and the openness of the building are integrated into an organic whole, realizing communication and exchange between the institute's innermost thoughts and the surrounding environment.

4

5

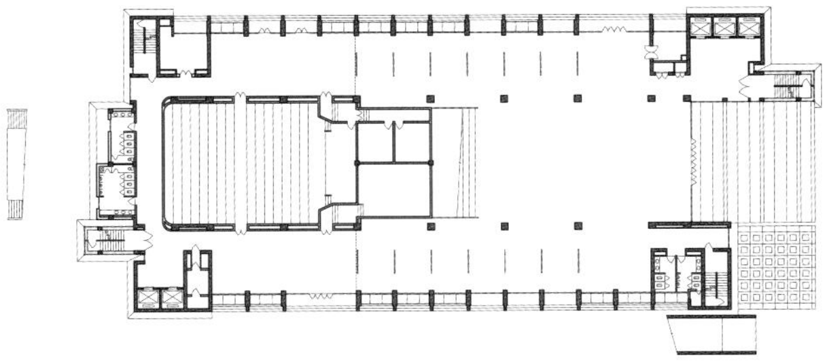

6

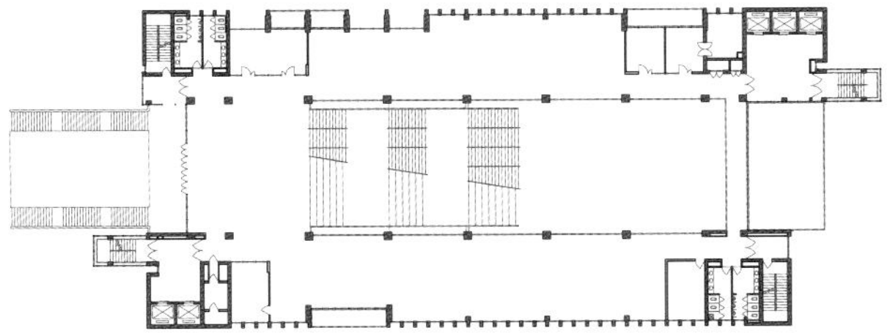

7

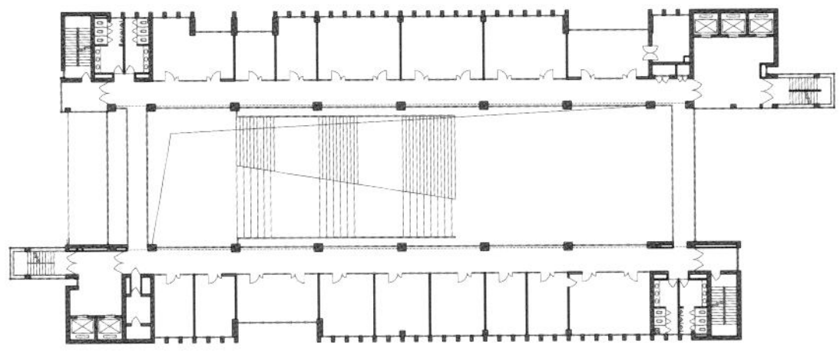

8

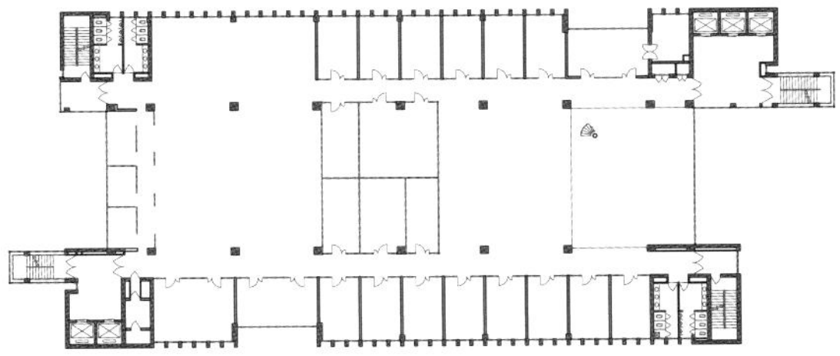

9

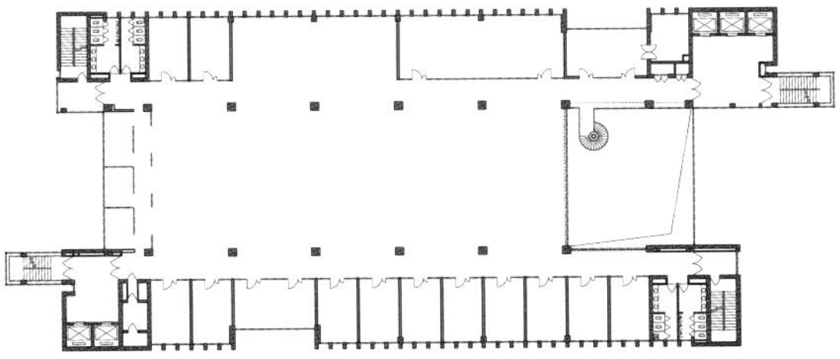

10

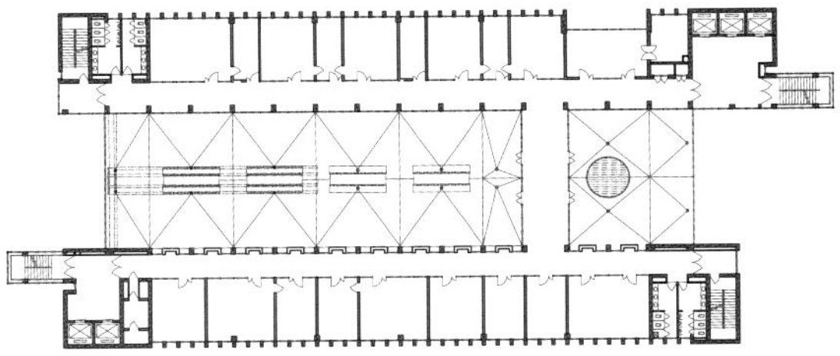

11

5 室内
Interior

6 首层平面图
The 1st floor plan

7 二层平面图
The 2nd floor plan

8 三层平面图
The 3rd floor plan

9 五层平面图
The 5th floor plan

10 六层平面图
The 6th floor plan

11 十一层平面图
The 11st floor plan

12

13

12 建筑外景 3
Exterior 3

13 建筑细部
Exterior detail

14 建筑外景 4
Exterior 4

14

中国科学院研究生院教学楼（中关村园区） 北京

Teaching Building of the Graduate University of CAS Beijing

设计时间 / Design：2001
建成时间 / Completion：2003
建筑面积 / Building area：65300m²
项目组 / Design team：崔彤 夏炜 王知非 王毓琳
业主 / Client：中国科学院研究生院
摄影 / Photographer：傅兴 杨超英

1 区位图
Location

2 建筑外景 1
Exterior 1

中科院研究生院第二教学园区，地处人文环境理想、科研文化浓厚的中关村基础科学园区，不足的是用地局促、限制较多。突出的问题是一期教学楼位于四面约束的狭长地段，南段的中科院发育所是主要影响因素，它曾以单元细胞式试验点荣获国家建筑设计金奖。业主希望保留此建筑，并成为将来教学楼的一部分，因此教学楼的设计以此作为出发点，并引发思考单元模式、对位关系、空间形态等诸多问题，形成三个单元串联的组织结构。

新旧之间的有机性并非是图形上简单的对位关系，而是教学单元可弹性生长的逻辑，既可以将原有的实验室改造为教学单元，并预留扩建通道的可能性，又是一个约束性法则，它规定着校园结构。一期作为〝现在时〞的话语权者，把过去和将来编织成一个秩序化的网络，它回答：过去可以〝那样〞；现在一定〝这样〞；将来可能〝怎样〞。每一部分都限制在规定的网格之中。

一期教学楼中合班教室和单班教室两种不同类型的并置是设计的主要矛盾，这导致了南北两列不同类型教学单元的布局。5.4m 高的合班大教室位于南端，3.6m 高的单班教室位于北端，中间自然形成了过渡空间〝内庭〞，空间的特质和趣味由此而引发。

南北边界的不平行迫使中庭协调矛盾，形成梯形街道空间，东边约 9m 宽，西边约 7m 宽，空间的缺陷也引发对空间透视变形的思考，期待空间纵向深度的表达，因此，设计中极端看重内庭所具有的矛盾性和复杂性，并希望在困境中寻求解答，比如，通过室内空间室外化，而使它具有一定的模糊性。本质上教学活动被覆盖在内庭之中，实际上街道空间所具有的自由感让学生自由地活动于一个没有约束的第二课堂。

The Second Teaching Area of the Graduate University of the Chinese Academy of Sciences is located in the Basic Science Park of Zhongguancun where there is an ideal human-oriented environment and a strong research culture, and the Park is home to various scientific research institutes. However, the disadvantage of this area is insufficient space and many restrictions. An outstanding problem is that the Teaching Building of Phase One is situated in the narrow strip with constraints on all four sides. The Institute of Genetics and Developmental Biology of the Chinese Academy of Sciences in the southern section is a main influencing factor, with the characteristics of a unit cell test unit. It was awarded the National Building Design Gold Award. The owners wanted to retain this building and use it as a part of the series of Teaching Buildings in the future. Therefore, the design of the Teaching Buildings takes it as a starting point, and triggers the concept of a unit model, a sequential relationship, spatial style and many other issues, forming the organizational structure in which the three units are connected with each other.

The link between the old and new building is not a simple sequential relationship on the graph, but rather, it is the logic of a teaching unit to be able to grow flexibly; it can either transform the original laboratory into teaching units, or be reserved for the possibility of future access expansion. However, it is a binding rule determining the campus structure. Phase One, as if it were the part of speech ˝in the present tense˝ , weaves the past and future into an orderly network：the past can be ˝like that˝ , and now it must be ˝like this˝ ; and ˝how˝ it may be like in the future. Each part is limited to the specified grid.

The arrangement of both single and joint classrooms in the Teaching Building of Phase One is the main conflict in the design, which leads to the different layout of teaching unit in the south and north directions. The 5.4m large joint classrooms are located at the southern end, and the 3.6m single classrooms are situated at the northern end, so the central area naturally forms an inner courtyard of a transitional space, inspiring nature and a taste of open space.

The unparallel north-south border forces the atrium to confront conflicts, the trapezoid street space is formed, and the width of the east side and the west side are about 9m and 7m respectively. The space defect also leads to the thinking of space perspective distortion, and the expression of vertical depth of the space is anticipated. Therefore, the contradiction and

2

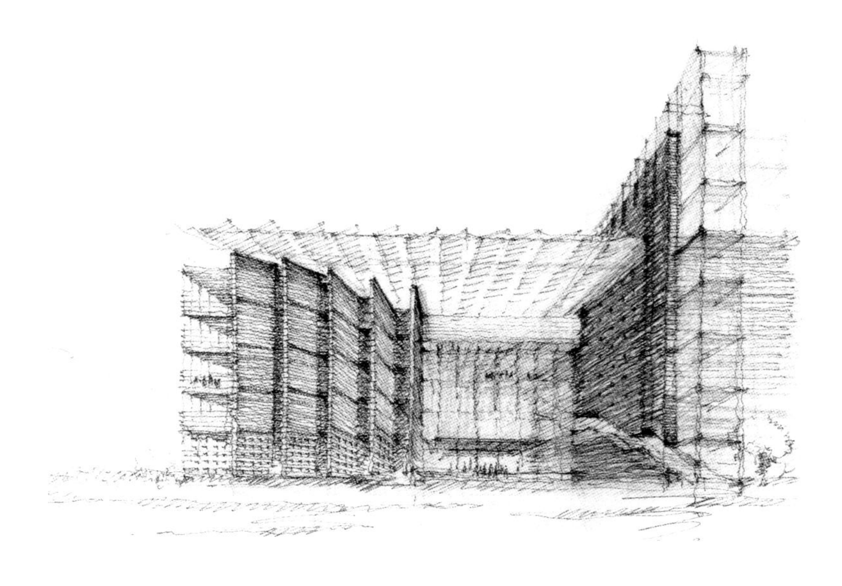

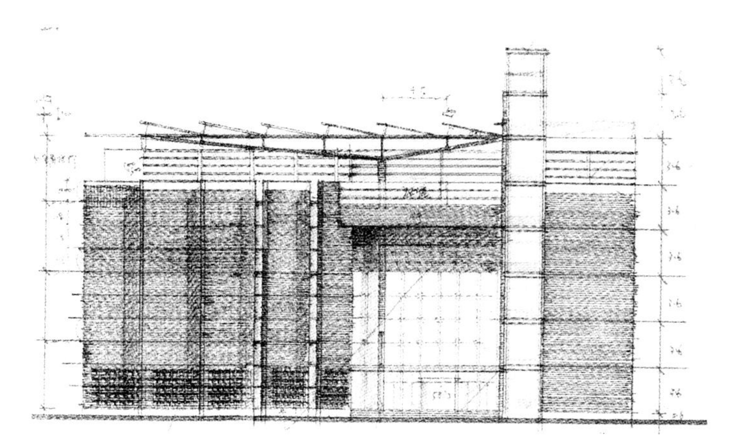

3

总之它的多重性是显而易见的，它不仅仅起着过渡和通道的基本作用，更提供了学生和老师见面、交谈和沟通的空间。不断地人潮流动好像编织了信息网络系统，内庭宛然一个知识的街道，成为这组建筑的主要命脉。

教学楼的主入口借助于中科院数、理、化、天、地、生六大学科关系中演变出六种渐变的墙面，从第一块垂直面到最后一块水平面，经历了四个墙面逐渐演化的过程，每片旋转 15° 角，直到最后一片变为水平。从量变到质变的数字式的变化，表面是一种渐开式的复合运动轨迹，包容着旋转和平移两个向量，但超越这层游戏般的规律，旨在寻找一个理性的思辨，并以婉转的态度诠释“引导”、“启发”和“循循善诱”的教育法则。多重信码的含义是把一本厚厚的书缓慢地打开，而又瞬间凝固，而这一动态语言是源于东方的含蓄精神，从潜藏于内部的力量由内而外地引爆——连同内庭的阳光和绿化、知识和信息一同乍泄。

complexity of the inner courtyard is a special concern in the design, and it is expected to solve the problem in the process. For instance, by making the interior space as exterior one, certain vagueness can be realized. The teaching activities are covered in the inner courtyard, and indeed, the freedom of the street space enables students to enjoy activities freely in a second classroom without constraints.

In short, the complex's multiplicity is obvious, and it not only plays an essential role of serving as transition and passage, but also provides a space for students and teachers to meet, talk and communicate. Continuous human flow seems to weave an information network system, and the inner courtyard, just like a street made up of knowledge, forms the main artery of this group of buildings.

By drawing on six academic disciplines of the Chinese Academy of Sciences, namely, mathematics, physics, chemistry, astronomy, geography and biology, the main entrance of the Teaching Building has six kinds of gradient walls. From the first vertical surface to the last horizontal surface, the four walls experience gradual changes, and each piece rotates 15 degrees until the last one becomes horizontal. Judging from the surface, the change of numbers from quantitative to quantification is a gradually opened composite motion, which includes two vectors of rotation and translation motion, but beyond this game-like rule, it is designed to find rational thinking, and interpret the ˝inductive˝, ˝inspiring˝ and ˝persuasive˝ educational law in an indirect manner. Multiple codes means to slowly open a thick book and keep it open, and this dynamic language originates from the spirit of inference in Asian countries, and it is set off by internal power—the sunlight, greenery, knowledge and information in the inner courtyard be revealed simultaneously.

4

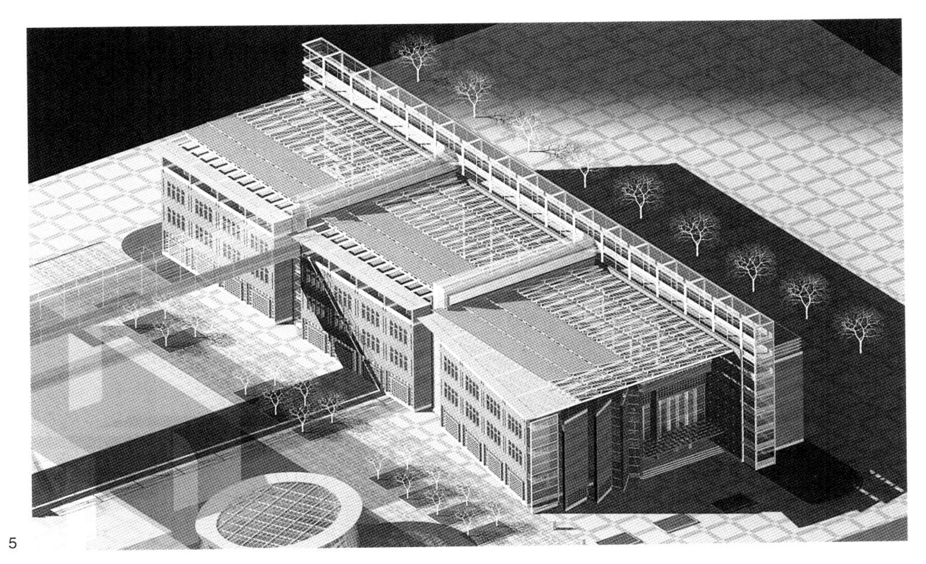

5

3 手稿
Sketchs

4 建筑外景 2
Exterior 2

5 模型
Model

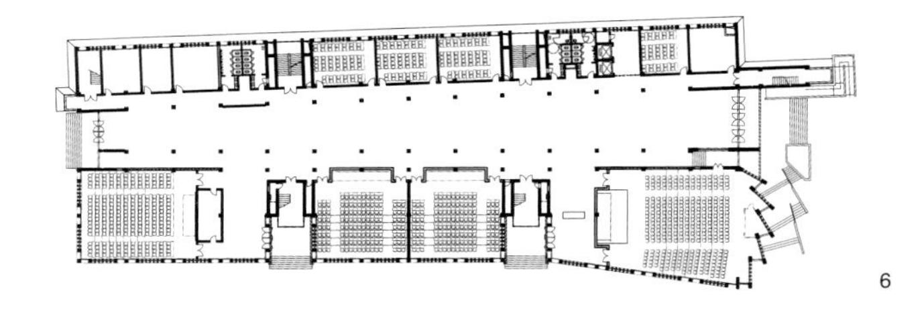

6

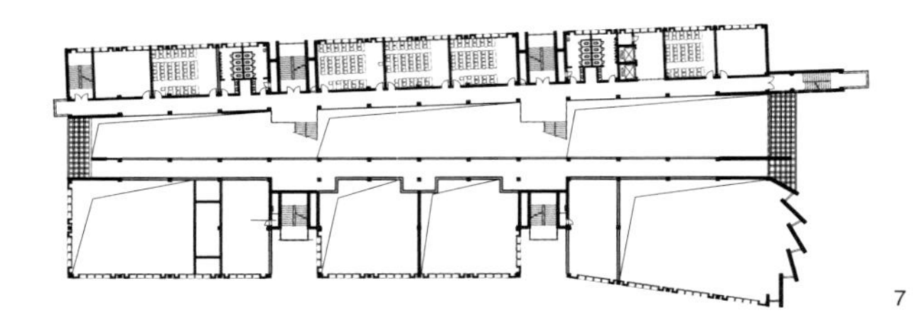

7

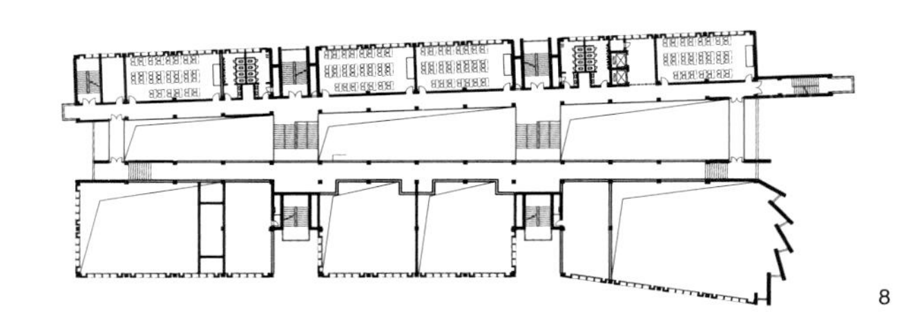

8

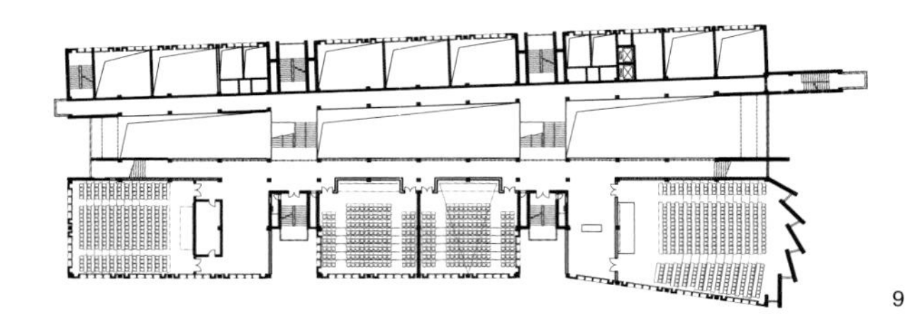

9

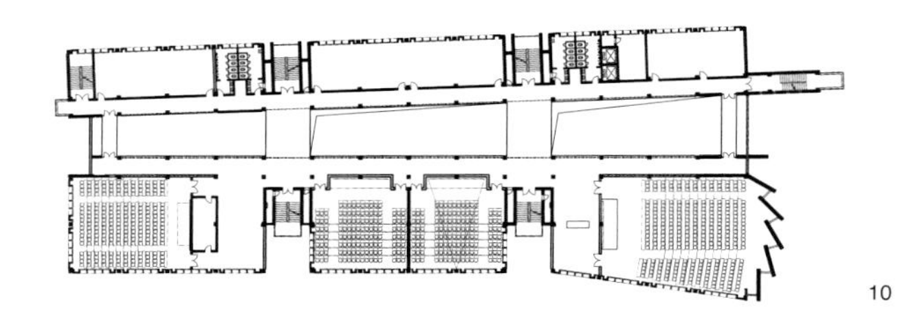

10

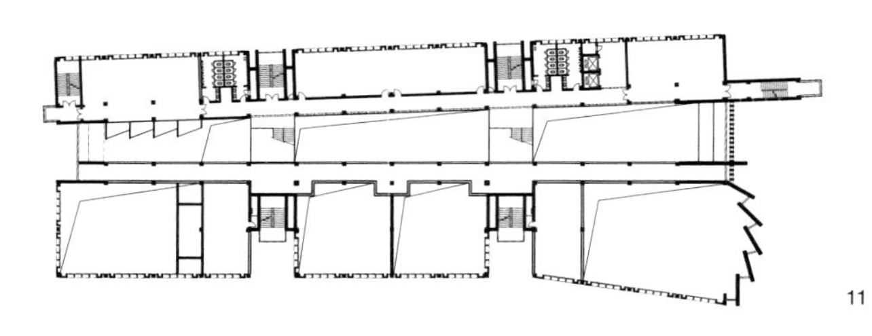

11

12

6 北段首层平面
North section
the 1st floor plan

7 北段二层平面
North section
the 2nd floor plan

8 北段三层平面
North section
the 3rd floor plan

9 南段二层平面
South section
the 2nd floor plan

10 北段四层平面、南段三层平面
North section the 4th floor, South section
the 3rd floor plan

11 北段五层平面
North section
the 5th floor plan

12 建筑室内 1
Interior 1

13 建筑室内 2
Interior 2

9

13

14

15

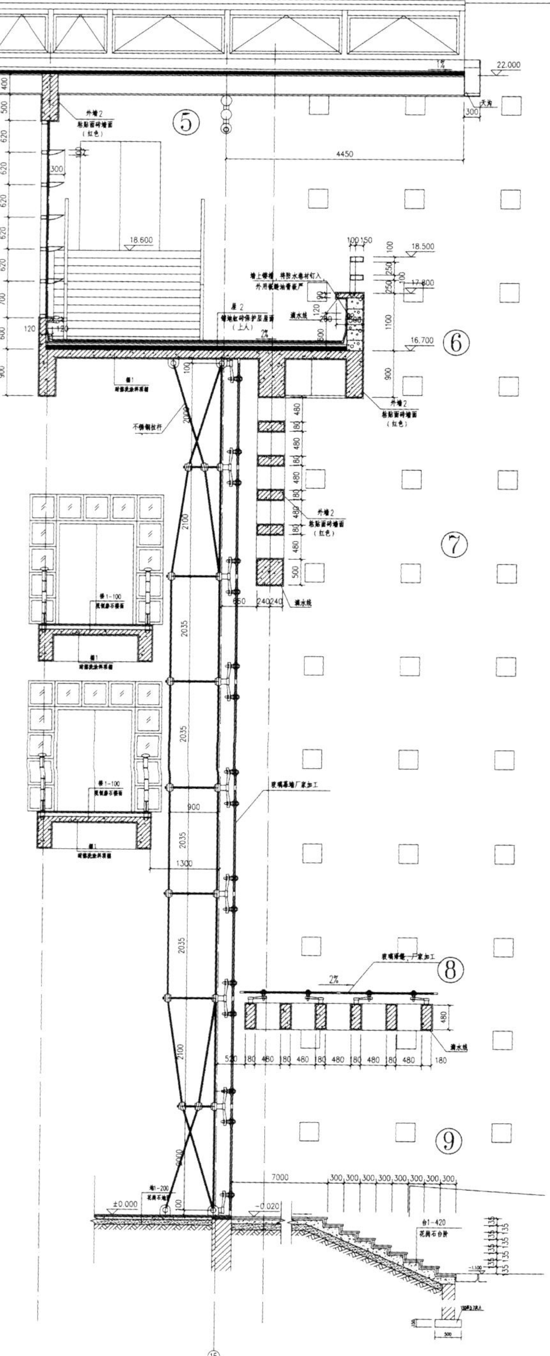

16

14 建筑室内 3
Interior 3

15 建筑室内 4
Interior 4

16 节点大样
Detail

17 建筑外景 3
Exterior 3

化学工业出版社　北京

Chemical Industry Press　Beijing

1

1　模型 1
Model 1

2　建筑外景 1
Exterior 1

设计时间 / Design：2004
建成时间 / Completion：2006
建筑面积 / Building area：11000m²
项目组 / Design team：崔彤 桂喆 何川 黄文龙
业主 / Client：化学工业出版社
摄影 / Photographer：杨超英

化工出版社厂房改造为办公楼，是制造业向知识产业的一种渐进式变革。机器"生产"的印刷品转化为思想"活动"编辑，存在着自相似性的"加工流程"。这一特殊性暗示活动行为微妙的转变，导致包裹"机器主体"的界面解体和包裹"人文主体"的形式重构，设计过程转变为"编辑加工"过程。新的立面与其说是设计不如说是改编，"巨构"的图书在条码的约束中重新再现旧工业化加工及工业化装配的过程，单一预制构件的排列组合装配出并不单一的产品。"新产品"源自工业生产加工，改造的精神则是"机器制造"和"厂房建造"的传承。

四个单元体块由于新功能的需求，被叠制、穿插和连通，功能分析图宛如一张空间构图。内部机能追随工艺美学逻辑之后，明显的目的是形成整体又相互独立的功能空间。墙面、水面、连廊、连桥自内而外，连续或间断、限定或流通，在简单而复杂的经验中理解密斯。代替两楼之间的吊车是一个桥或门洞，暗示连通和流动，并将公园的景色穿过桥梁"嵌入"其中，最终的目标是将建筑"锚固"于公园之内。建筑和环境彼此的联系是一种形而上的对话，在流动空间中蜕变出中国园林的意境。

混凝土的魅力铸造了一种永恒。混凝土的品质从朴素平实到庄重权威，似乎无所不能，它具备无限的宽容度和适宜性。在新与旧、人工与自然之间搭建起一座桥梁。垛斧之后"风化"般的质感又仿佛回到原点，让我们记起火山的浮尘熔炼一种精神，它曾改变了一个世界，构筑了工业时代机器般的铿锵。此时，混凝土在机器的压痕中重塑一种"场所精神"。

厂房留下工业化生产的建造过程，被意识到潜藏的价值。立面细节中尝试另类的工业化体系，介于标准与非标准之间的组织系统，在南、北、西三面做出反应。不同于习惯性的做法是，开敞的北面极限状态地呼吸空气，探求缓冲层表皮。改造的趣味好像是破坏，其实是继承。工业化的语言又一次演变出新的模数化格式，以灯芯绒混凝土预制板牵引出一股粗鲁的诗意，并玩笑似的披上童话般的外衣——森林中的书屋。无论是森林中的书屋，还是书屋中的森林，"改"变后的建筑再"造"了第二自然。

The renovation from a plant to an office building for Chemical Industry Press symbolizes the subtle revolution of shifting the focus from manufacturing to a knowledge industry. In the past, we paid more attention to manufacturing more presswork with machines; but now, the focus is on the editing of "thoughts"; both of them have similar work flows. This conclusion implies the subtle change of actions, which leads to the disintegration of the main body that was the "machine" and the reconstruction of the main body which is now "humanity", as well as the change from the design process to the "editing" process as well. The new elevation, rather than being a new design, is a reform of the old one. The giant book-shape structure represents the industrial work and assembly process. Although the new structure is made up of prefabricated parts of a single size, it is not a monotone structure owing to careful arrangement and combination. The "new product" originates from industrial production processes, and the renovated spirit is the heritage of "machine manufacturing" and "plant construction".

The four units are interweaved and interconnected to meet the demands of the new functions. The functional analysis diagram is just like a space configuration. The internal function aims at forming an integrated space with independent functional units, while conforming to aesthetic tastes. The wall, the water surface, the vestibule and the platform bridge in the building form an integrated whole that are continuous or disconnected, limited or unblocked, and embody both simplicity and complexity. Instead of a crane, a bridge or a door opening is arranged between two buildings, symbolizing the intercommunication between the two buildings. Meanwhile, this creative design brings the scenery of the garden into people's view. As a result, it seems that the building is built in the midst of the garden. The metaphysical connection between the building and the environment creates the artistic style of the traditional Chinese garden.

The enchantment of the concrete structure symbolizes eternity. Concrete represents simplicity, gravity and authority. It is all-inclusive and tolerant. It helps to bridge the new and the old age, artificial construction and natural environment. Its "weathered" texture reminds us of the spirit of a volcano spewing lava and dust. It changed a world and built a forceful industrial time. At this moment, concrete reproduces a sort of "site spirit" under machines.

The hidden value of the industrial processing procedure is well-recognized. The elevation detail attempts to set up a different industrial system which is expected to be an intermediate one between the standard and the non-standard order. It tries to design the south, west and north sides of the structure in different manners. Different from common practice, the north side is created to be an open space to take in as much fresh air as possible and explores the surface of the buffer layer. It reforms rather than destroys the original building, but inherits the merits of the old construction. The old building of the industrial age is endowed with new styles. The ribbed concrete pre-cast plate is both rough and poetic; the combination of the woods and the 'book-shape house' invites us into a fairy tale. Be it a book-shape house in the woods or the woods in the book-shape house, the renovated building creates a second nature.

2

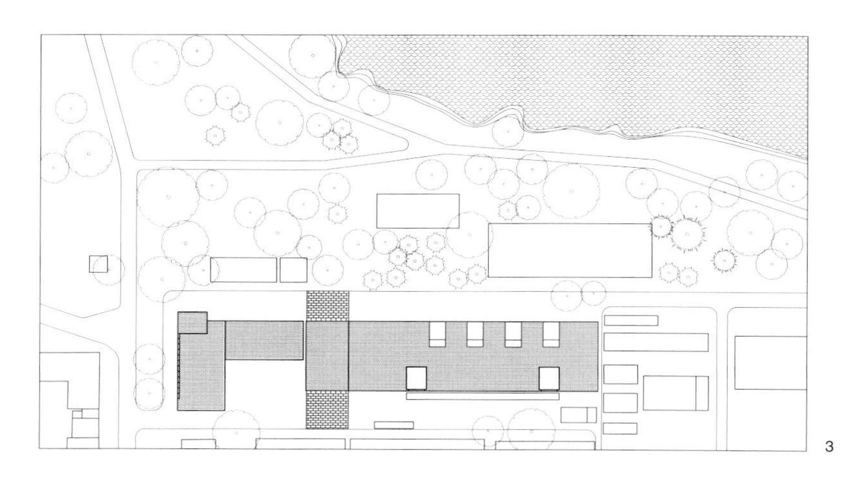

3

3 总平面图
Site plan

4 建筑外景 2
Exterior 2

5 解析模型
Analytic model

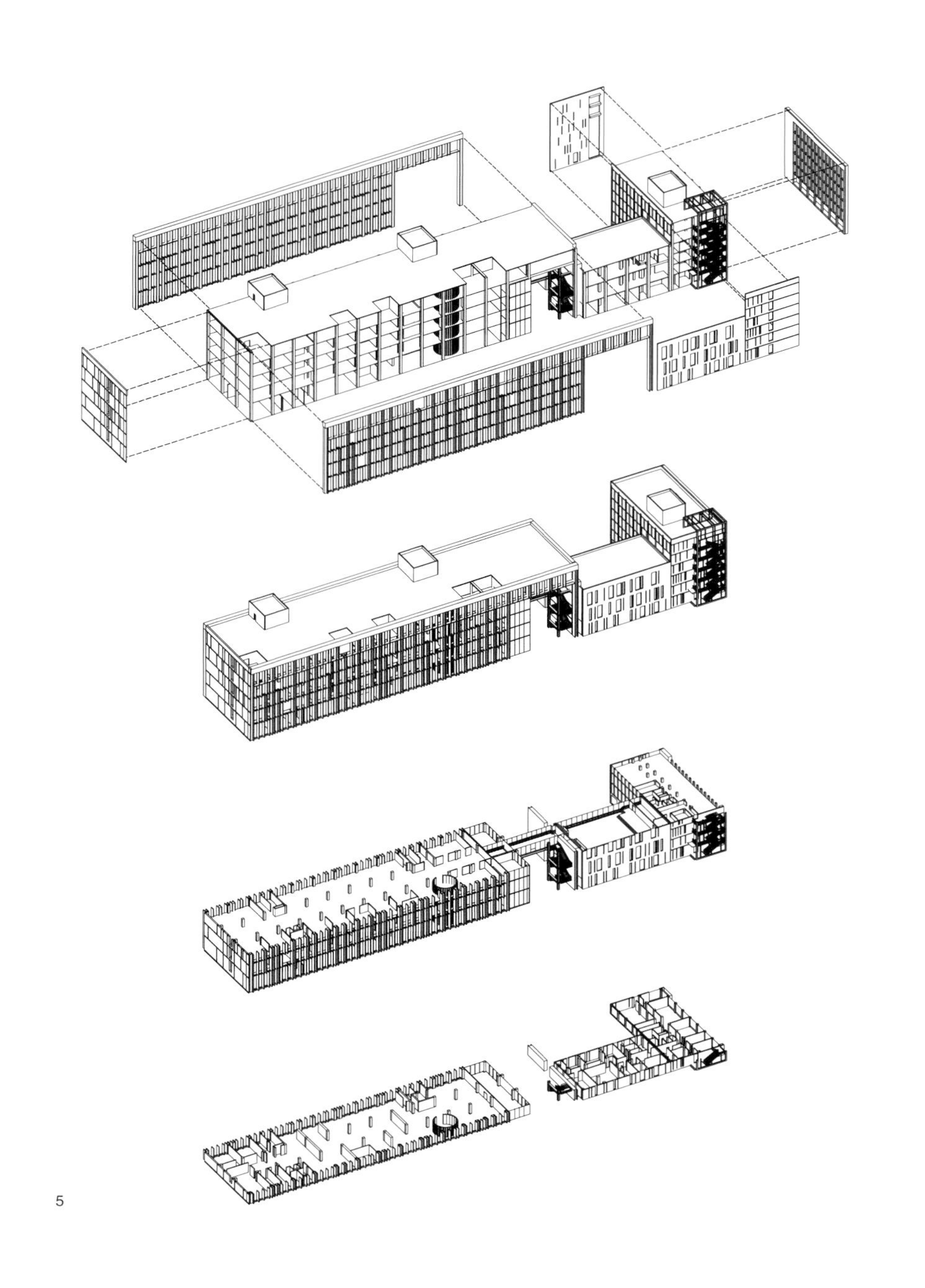

5

4

6 建筑外景 3
Exterior 3

7 建筑外景 4
Exterior 4

8 建筑外景 5
Exterior 5

9 建筑外景 6
Exterior 6

6	7
8	

9

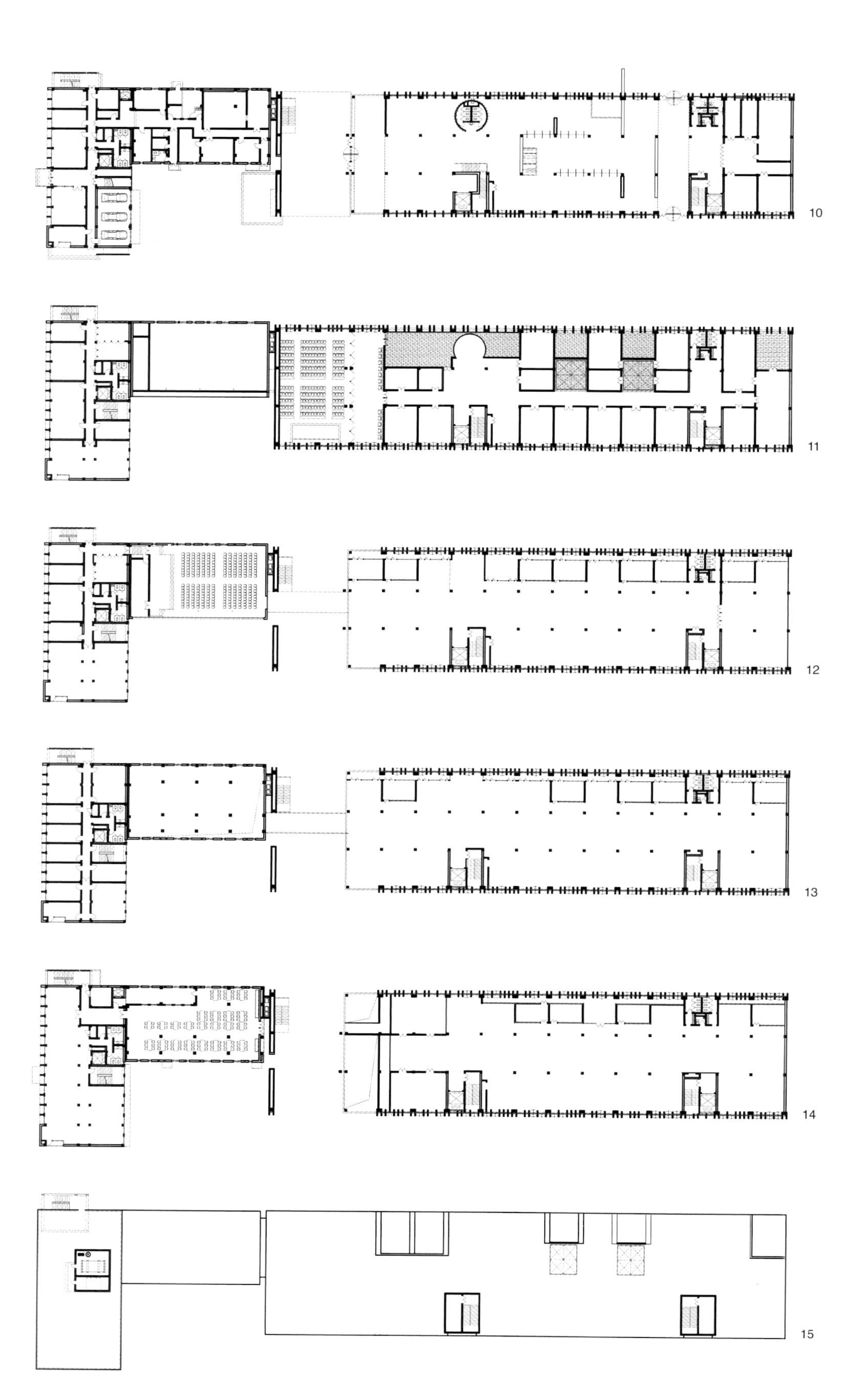

10 首层平面
The 1st floor plan

11 二层平面
The 2nd floor plan

12 三层平面
The 3rd floor plan

13 四层平面
The 4th floor plan

14 五层平面
The 5th floor plan

15 六层平面
The 6th floor plan

16 建筑外景 7
Exterior 7

16

17

17 建筑外景 8
Exterior 8

18 建筑外景 9
Exterior 9

19 建筑外景 10
Exterior 10

18

19

国家动物博物馆·中国科学院动物研究所 北京

National Zoological Museum of China / Institute of Zoology, CAS Beijing

设计时间 / Design：2002
建成时间 / Completion：2006
建筑面积 / Building area：$42900m^2$
项目组 / Design team：崔彤 桂喆 何川 白小菁 文业清
业主 / Client：中国科学院动物研究所
摄影 / Photographer：舒赫 杨超英

1 区位图
Location

2 建筑细部 1
Exterior detail 1

1

位于奥体公园西、奥运村南的中国科学院天文、地理、生命园区是国家级科学实验园区，作为科学共同体的最大园区——动物研究所，从周边环境中吸取若干属性，并转化为具有独特科学内涵的科研群体。

动物研究所总体布局的新模式在科研、教育、成果转化的平台基础上，继续向复合化迈进，体标本馆作为相对独立的储存和保护的专业馆；对外展览和普及教育的国家动物博物馆自成体系。总体布局依据不同的实验单元和公共空间，确立研究所为复合园区。

开放体系

研究所的开放体系首先意味着从科学神坛走向民众，成为城市的活力生发点，为该地区提供科学的"能量"。其二，无墙的园区促进了各研究所及各学科之间的交叉和资源共享，动物研究所将成为内外能量交换和新陈代谢的"生命体"。

场所感的建筑不仅是此时此地的建筑，也是那时此地的建筑，它要传承过去、关注现在、面向未来。场所感的建筑要以最小限度地影响环境为原则，以嵌入式的空间融入到环境中，并成为场所的有机体。探索生命科学内涵，为科学家创造静谧、理性、儒雅的科研空间成为设计的出发点。

动物所由 9 个单元集成为系统的空间结构，并遵循有机生长和衍生逻辑。这组动态均衡的园区中，有一根主轴线统领形成纵深发展的空间序列。主轴线起始于南侧博物馆前广场，发展于研究所三合院的外向型广场，通过主门庭延伸至北端廊院花园，终结于生殖楼的内庭院。内外广场的过渡体现在南部由动物博物馆和地理博物馆形成的社会性广场与动物所园区南端主广场的联系而形成的复合广场。穿越南端复合广场的第二个层次是廊院花园，与大园区的花园沟通，形成连续的绿野景观；第三个层次是北端生殖楼的庭院嵌入到大花园中。

建筑形态

骨骼：建筑形态是内在结构和功能的反映，建筑结构如同骨骼，作为支撑体系在关注单元体建构中更强调单元之间的过渡和联系，如同"关节"的传承作用。

关节：关节作为连接点，将各单元组合为完整的骨架系统，预示着生长的可能性，自身机能的特殊性则体现在两个不同单元之间的微妙过渡，如门庭和交通核心或公共空间。

The astronomy, geography and life park of the Chinese Academy of Sciences, located west of the Olympic Park and south of the Olympic Village, is a national-level scientific experiment park. As the largest park in the scientific community, the Institute of Zoology draws several attributes from the surrounding environment to transform it into a scientific research complex with unique scientific intentions.

In the general layout, with the platforms of scientific research, education, and achievements, the Institute of Zoology continues to pursue the design of a complex. For example, the specimen hall as a relatively independent specialized hall for storage and protection, establishing its own system and houses the National Zoological Museum of China for exhibition, public use and education. The general layout establishes the Institute as a complex park depending on different experiment units and public spaces.

Open system

Firstly, the open system of the Institute means moving towards the people from scientific seclusion to become the source of vitality that is the growing point of the city and offers scientific "energy" for this area. Secondly, the wall-free park promotes resource sharing among the research institutes and disciplines. The Institute of Zoology will become the "life entity" for the exchange of internal and external energy and growth.

The Institute of Zoology is a systematic structure of space integrated by 9 units, which follow the growth of organisms and the source of logic. In this dynamic and balanced park, the main axis demands the need to form a space developed in depth. The main axis originates from the square in front of the museum at the southern side, developing through the open square in the three-section compound of the Institute, extending to the corridor courtyard garden in the north via the main courtyard, and ends in the interior courtyard of the Reproduction Building. In the south, the transition between the internal and external squares embodies the connection between the social square formed by the Zoological Museum and the Geological Museum and the southern main square of the Institute's park, thus forming a square compound. At the second level, the corridor courtyard garden passes through the south compound square which links with the garden of the large park, forming a continuous green field view; and at the third level, the courtyard of the Reproduction Building in the North is embedded in the large garden.

Architecture form

Skeleton：Architectural form is the reflection of inherent structure and

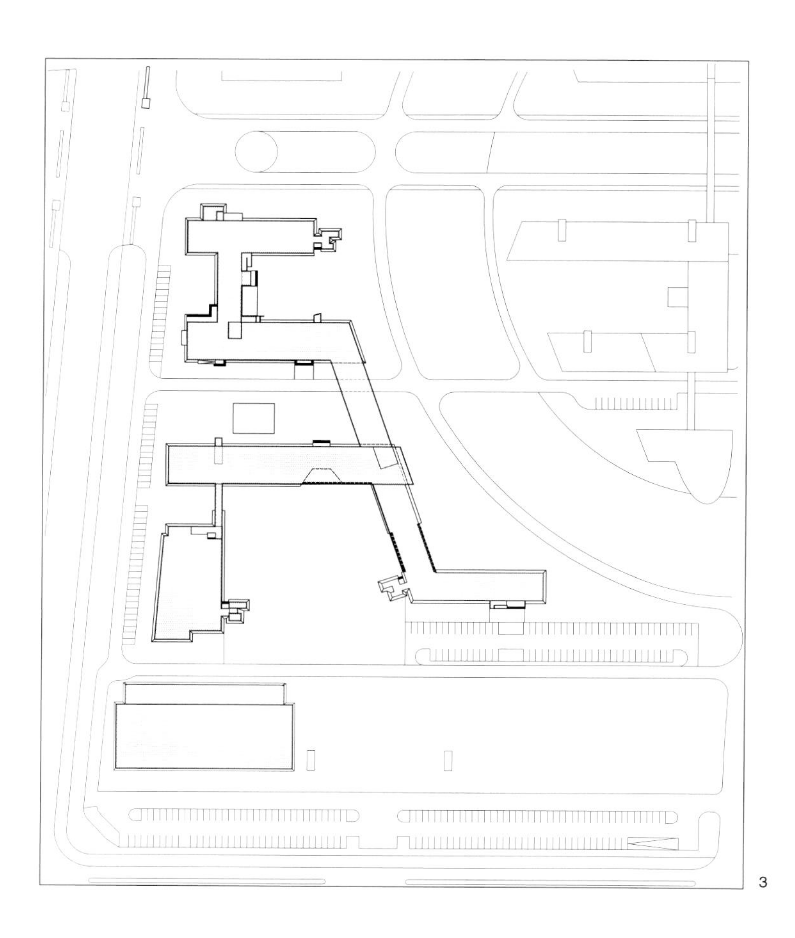

3 建筑总图
Site Plan

a 科研楼
Institute
b 动物馆
Museum

表皮：建筑的界面作为有意味的表皮，是骨骼系统和功能逻辑的真实体现，在不同的单元中呈现出相同的皮肤和不同的肌理，比如标本馆的空腔实体间层所形成的生物气候缓冲层，确保标本的恒温、恒湿。

模数：在均值变化中形成一个多层次的模数关系，以求石材单元划分与开窗的协调统一，并兼顾采光、通风、遮阳系统的完整。

DNA：国家动物博物馆的形态和而不同，是对大空间博览建筑的真实还原。它借助仿生学应用连续的排架结构，建构了一种渐变韵律的宏大空间，作为一种类骨架系统，在展示自然有机生物标本的同时，重构了一种新的DNA有机建筑。运动和扭转的博物馆在融入生命园区中，以一种新语言创造了通向奥运场馆的新景观。

functions. Like the skeleton, the structure is the supporting system which pays more attention to transition and connection between the units in the architectural construction, functioning as "joints".

Joints: As the connection points, the joints integrate each unit into a complete skeletal system, and indicate the possibility of growth. Meanwhile, its particular functions are reflected in the delicate transition between two different units, such as the courtyard and traffic core or public space.

Epidermis: The interface of buildings is like epidermis providing the true embodiment of skeletal system and functional logic and presenting the same skin and different skin textures in different units. For example, the bio-climate buffer layer formed by the cavity entity of the specimen hall ensures the constant temperature and humidity for the specimens.

Module: The epidermis unit for standard scientific research forms a multi-level module relationship in terms of changes in the mean value, so that the classification of a stone unit can meet the coordination and uniform requirement for window opening and the completion of lighting, ventilation, and shading systems.

DNA: The National Zoological Museum is harmonious but different in form, as it is a true restoration of the large space museum building. Because of its study of life sciences, the Museum applies a bent structure to construct a grand space in a gradual, regulated transformation. As a kind of frame-like system, it shows biological specimens and also reconstructs a new DNA organic building. By incorporating into the life park, the dynamic and reversed museum creates a new view leading to the Olympic venues in a new language.

4

4 建筑外景 1
Exterior 1

INSTITUTE OF ZOOLOGY CAS
國家动物博
國家动物博物館
中国科学院动物研究所
中国科学院

5

5 建筑内院 1
Courtyard 1

6

6 建筑外景 2
Exterior 2

7 博物馆、科研楼首层平面
The 1st floor plan of museum and research building

8 三层平面
The 3rd floor plan

9 科研楼、标本楼五层平面
The 5th floor plan of research building and specimens building

10 建筑内院 2
Courtyard 2

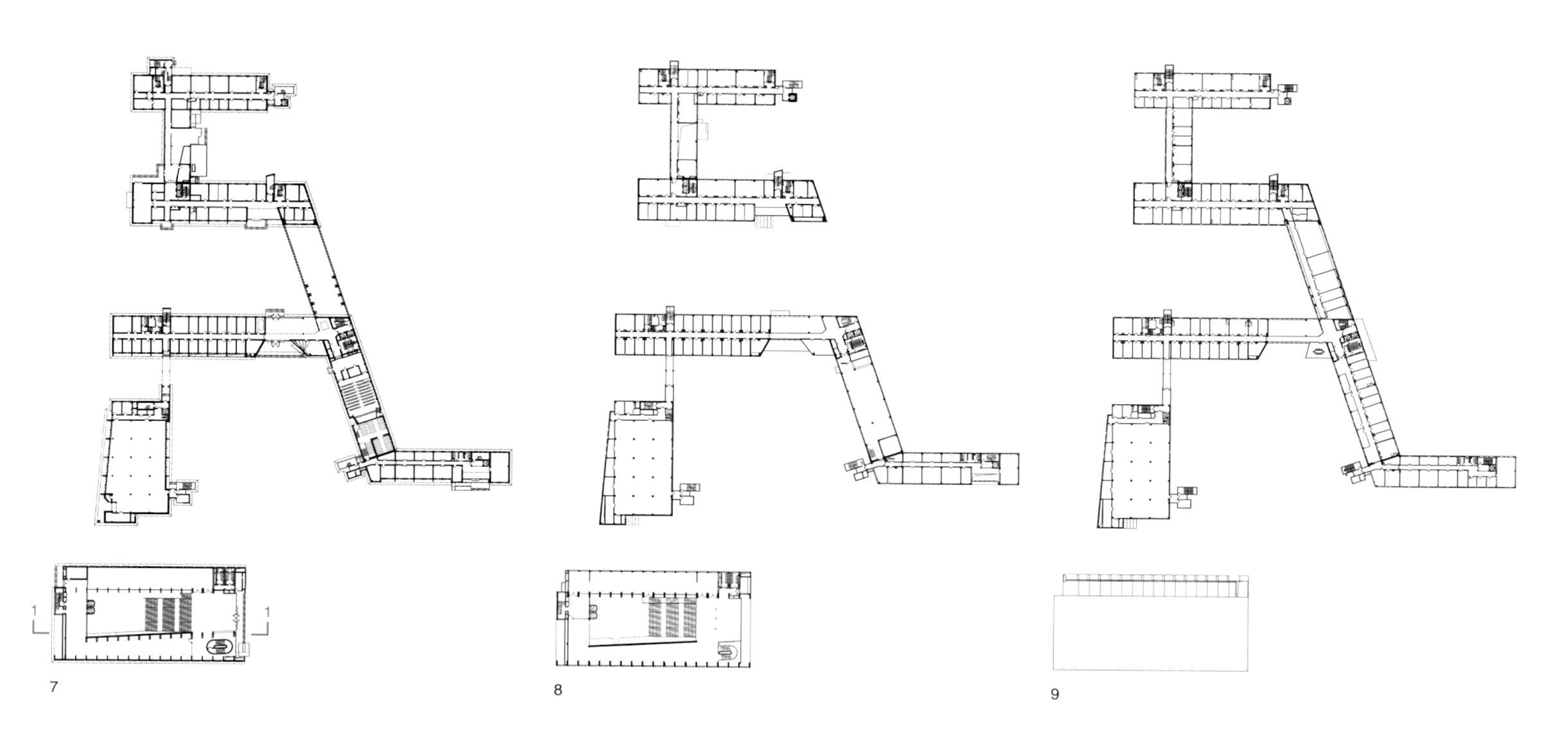

7

8

9

10

11 | 12
13 | 14

15

11 建筑外景 3
Exterior 3

12 建筑外景 4
Exterior 4

13 建筑内院 3
Courtyard 3

14 建筑外景 5
Exterior 5

15 建筑内院 4
Courtyard 4

16

16 建筑室内
Interior

17 建筑外景 6
Exterior 6

18 建筑外景 7
Exterior 7

17

18

721 国家重点工程 北京

721 National Key Project Beijing

设计时间 / Design：2006
建成时间 / Completion：2010
建筑面积 / Building area：52000m²
项目组 / Design team：崔彤 刘向志 于鹏 桂喆 韩冰洁
摄影 / Photographer：杨超英

1 区位图
Location

2 建筑外景 1
Exterior 1

作为国家级金融涉密工程，在满足"中国的、金融的、园林的"同时，最为特殊的要求是："品质卓越、印象淡泊"。面对这样一个挑战性难题，设计策略是"融"、"隐"、"弱"，但并不是意味着让建筑"消失"，相反是让建筑身份和场所发生最本质的相关性。

设计的目标是基于中国传统的"二元中和"思想完成一个隐没于城市中的山水空间的重构。

选址：本案选址于北京内城区与西山脚下之间的一块交通便利的场地，以满足都市的便捷和自然的静谧的双重要求。

规划布局：中心主轴线控制着核心区，以理性的秩序决定着特殊功能的"礼"；东边宛转和自由的笔法引导者园林化的辅助空间。最终形成理性与感性交织的统一体。

功能解析：核心功能单元以高敞、大跨度空间为主导并适应于人、物、智能化机器人的移动、装卸和守护的特殊使命。"安全、坚固、保密"是关键，"安保工艺"决定"三重功能层级"，形成以"山丘"围合下的"外向型"花园，内向型单元和静谧型核心区套层体系。由"外"、"内"、"核"的对立统一构成一个特殊的"城"。

建筑形态：建筑形态是"大小、黑白、图底、虚实、简繁"二元关系的经营和运筹。建筑化整为零后又化零为整，形成分散又连续的"消解式"体量；将主体功能单元立面处理为均质的中国格栅式背景，以脉动起伏的黛瓦屋顶提示公共空间入口。

主体建筑的大面积留白及简洁的方格母题为背景，创造出一个半透明的巨型宣纸，点缀几分墨色，写意出山水园林景象。

As a national-level financial project, when meeting the concepts of "China, Finance and Garden", the most special requirement is："transcendent quality, indifferent impression". Facing such a challenging puzzle, the design strategy is "melt", "conceal", and "weaken". While this does not mean letting the building "disappear", on the contrary, to generate the most essential correlation between the identity and place of the building.

The goal of design is to reconstruct a landscape space concealed in the city, on the basis of traditional Chinese "dual neutralization" ideology.

Site selection：The site of this case is selected in a field with convenient transportation between the city of Beijing and the Western Hills, to satisfy the dual requirements for metropolitan convenience and natural serenity.

Programming and layout：The principal central axis controls the core area, and determines the "ritual" of special functions in rational order. In the east, insinuating and free drawing leads the auxiliary space of the afforestation. Ultimately, a unity integrating rationality with sensibility is formed.

Function analysis：The key function unit is dominated by high, spacious and large-span space, and also applied to special missions of movements, loading, unloading and guardian of human, object and intelligent robots. "security, solid, and confidentiality" are the key, "security craft" determines. The "tier of triple functions", and forms the layered system surrounded by "hills", including "outward" gardens, inward units, and serene core areas. A special "city" is formed by the unity of opposites of "outward", "inward" and "core".

Form of building：The form of the building is the management and planning of the dual relations among "big or small, black or white, figure or ground, false or true, simple or complicated". The whole building is broken up into parts, which are then assembled the parts into whole, thereby forming a disperse and continuous "digestion" volume, the elevations of main functional units are processed into homogeneous Chinese grid-type background, and the entrance of public space is prompted by black roof-tiles that are up-and-down.

A large area of the main body of the building remains blank, with brief grid motif taken as the background. A piece of semitransparent giant rice paper is created to embellish a bit ink color, and landscape and garden scenery is drawn.

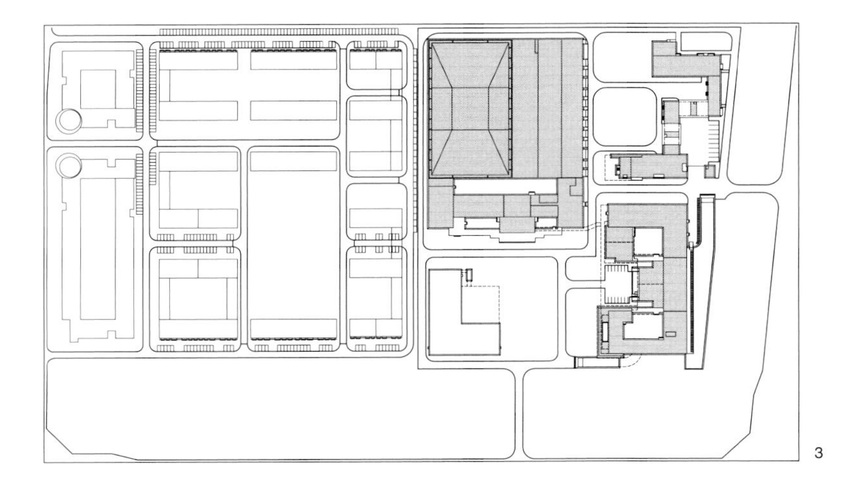

3

3 总平面图
Site plan

4 建筑外景 2
Exterior 2

5 解析模型
Analytic model

4

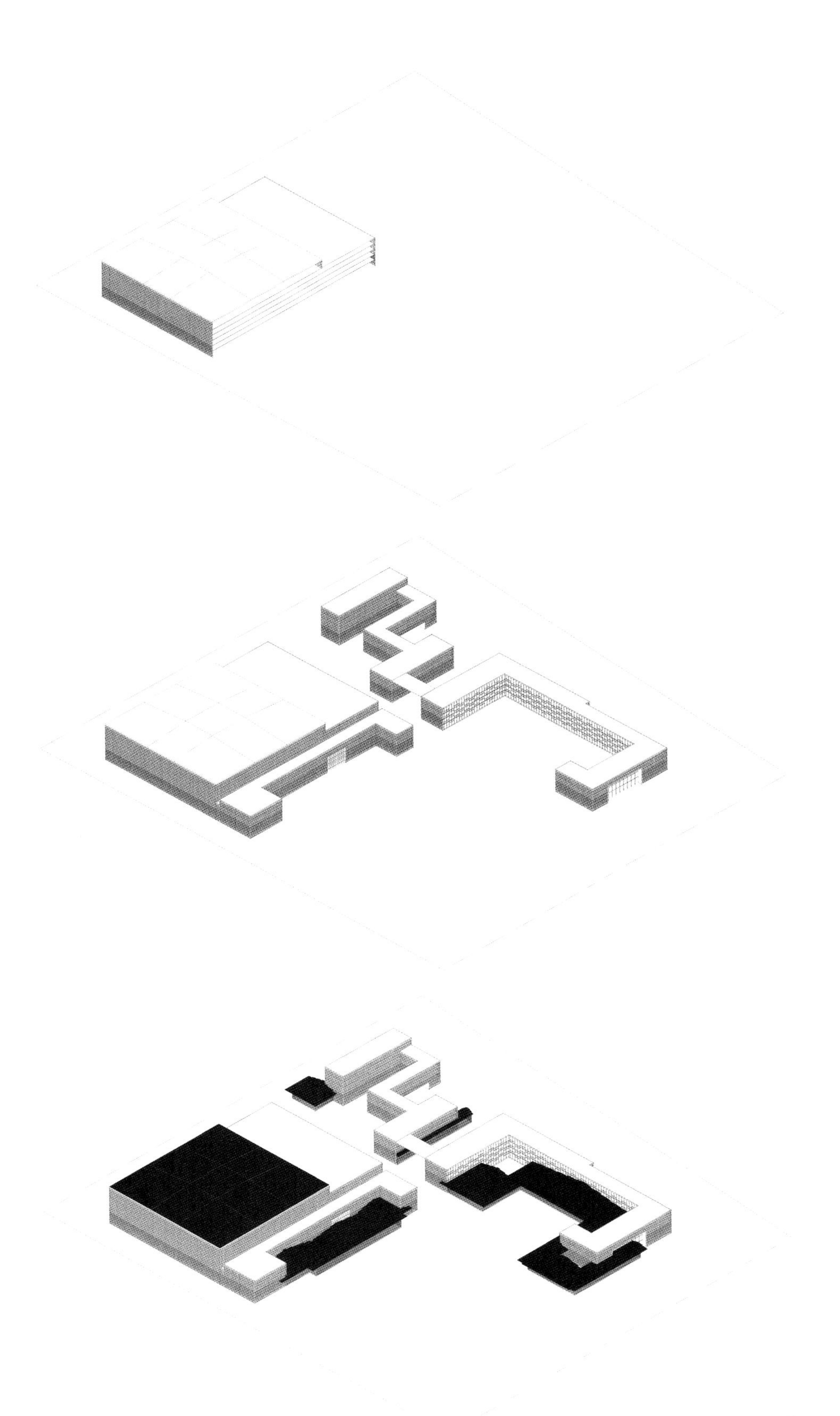

5

6

6 建筑外景 3
Exterior 3

7

8

9

7 建筑庭院 1
Courtyard 1

8 建筑庭院 2
Courtyard 2

9 建筑细部
Exterior details

10

10 建筑外景 4
Exterior 4

北京太伟高尔夫俱乐部 北京

Beijing Taiwei Golf Club Beijing

1

设计时间 / Design：2002
建成时间 / Completion：2004
建筑面积 / Building area：35000m^2
项目组 / Design team：崔彤 赵正雄 曾荣 华夫荣
业主 / Client：北京太伟集团
摄影 / Photographer：杨超英 傅兴

1 模型
Model

2 建筑外景 1
Exterior 1

山地俱乐部设计源于〝地脉〞文化的思考，努力创造符合高尔夫运动特征的建筑形态；营造一组休闲、健康、个体鲜明的室内外空间；构筑建筑、山地、林木有机相融的生态场所。

区别于城市建筑的理性构思是对〝规矩〞的重新认识，地形、山势、林木成为基准点并引发思考：脉动、包容、留白、移栽、转折、过渡、错位等新的建筑语言，因此，自由的呼吸、自由的伸展成为建筑秩序的新目标。

犹如山地起伏的建筑轮廓线与其说是山的写意，毋宁说是建筑功能逻辑下的形式反映。中庭下面所覆盖的正是山地中保留下的重要元素，作为建筑的一部分同时包含着〝山石〞、〝植被〞的自然属性，并与建筑相互依存，成为场所中的环境因子。

建筑能够作为第二种自然，不仅在于木百叶和粗野主义遗风下的〝灯芯绒〞混凝土的表现，而且在于对山地肌理的执着追求，最终的结果是潜藏于体内的〝爆发力〞转化为山地中的运动建筑。

Originated from the thinking of 〝geographical artery〞 culture, the design for the Beijing Taiwei Golf Club attempts to create one architectural form to confirm the sport golf's features：a group of interior and exterior spaces for leisure and health, and an ecological venue harmoniously incorporating the buildings, hills and woods.

A rational thinking different from urban building design requires a new understanding of terrain, hills and woods, and relies on them as benchmark to reconsider architectural languages such as fluctuation, containment, reserving vacancy, transplanting, conversion, transition, and transposition. Therefore, free breathing and stretching becomes a new target of architectural order.

The building's profile as a series of undulating uplands is a reflection of form under the building's functions, rather than a painting of hills. The area covered by the atrium is a major element of the hills, such as stones, vegetation and other natural attributes as one part of the architecture. These natural elements coexist with the architecture and become environmental factors of the venue.

Being a kind of second nature, the architecture is not only a presentation of timber blinds and brutalism of 〝corduroy〞 concrete, but also a constant pursuit of arteries and texture of the natural geological features. The final outcome is a dynamic piece of architecture on hills transformed from the 〝explosive force〞 hidden in the body of nature.

3

4

5

6

3 建筑外景 2
Exterior 2

4 手稿
Sketch

5 总平面图
Site plan

6 建筑外景 3
Exterior 3

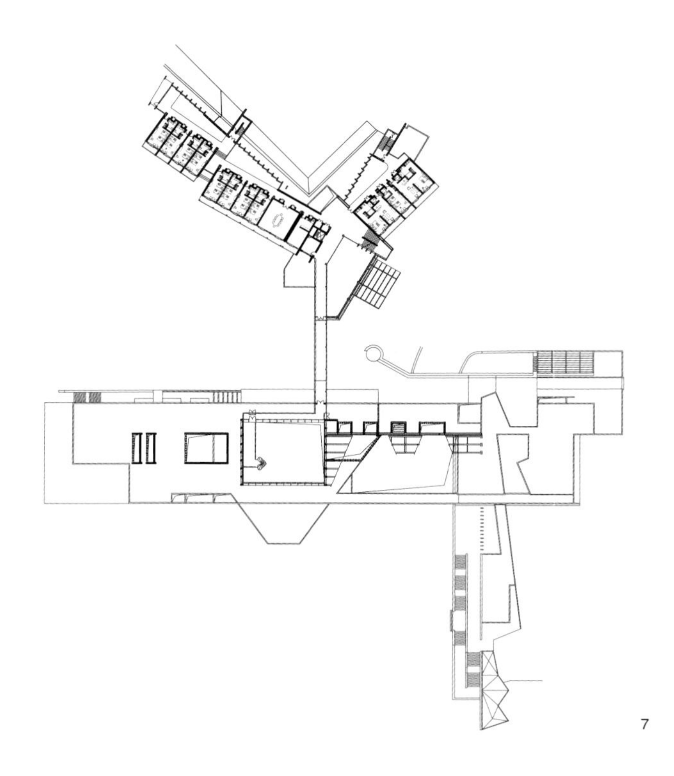

7

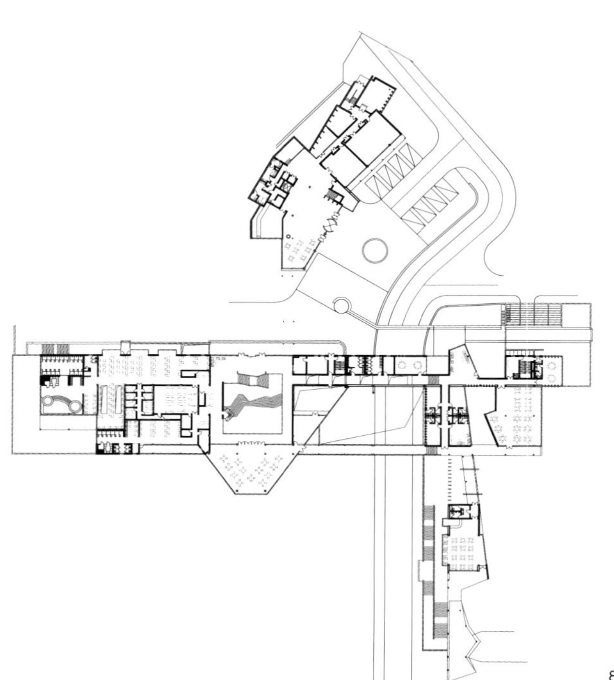

8

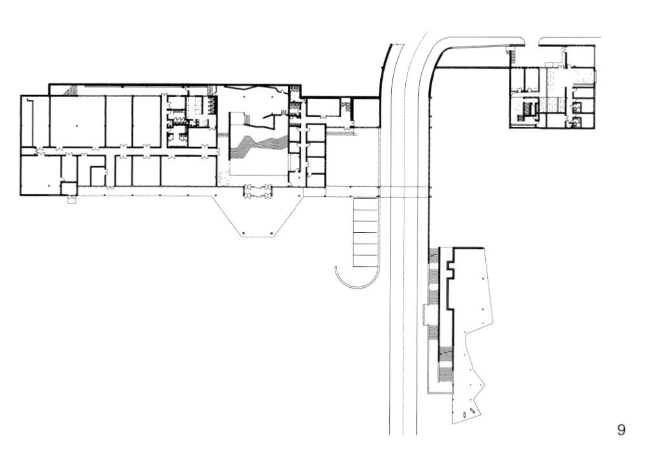

9

7 三层平面图
Third floor

8 二层平面图
Second floor

9 首层平面图
First floor

10 建筑外景 4
Exterior 4

11 建筑外景 5
Exterior 5

12 建筑外景 6
Exterior 3

10

11

12

13

13 建筑细部 1
Exterior detail 1

14 建筑细部 2
Exterior detail 2

15 立面图
Elevation

14

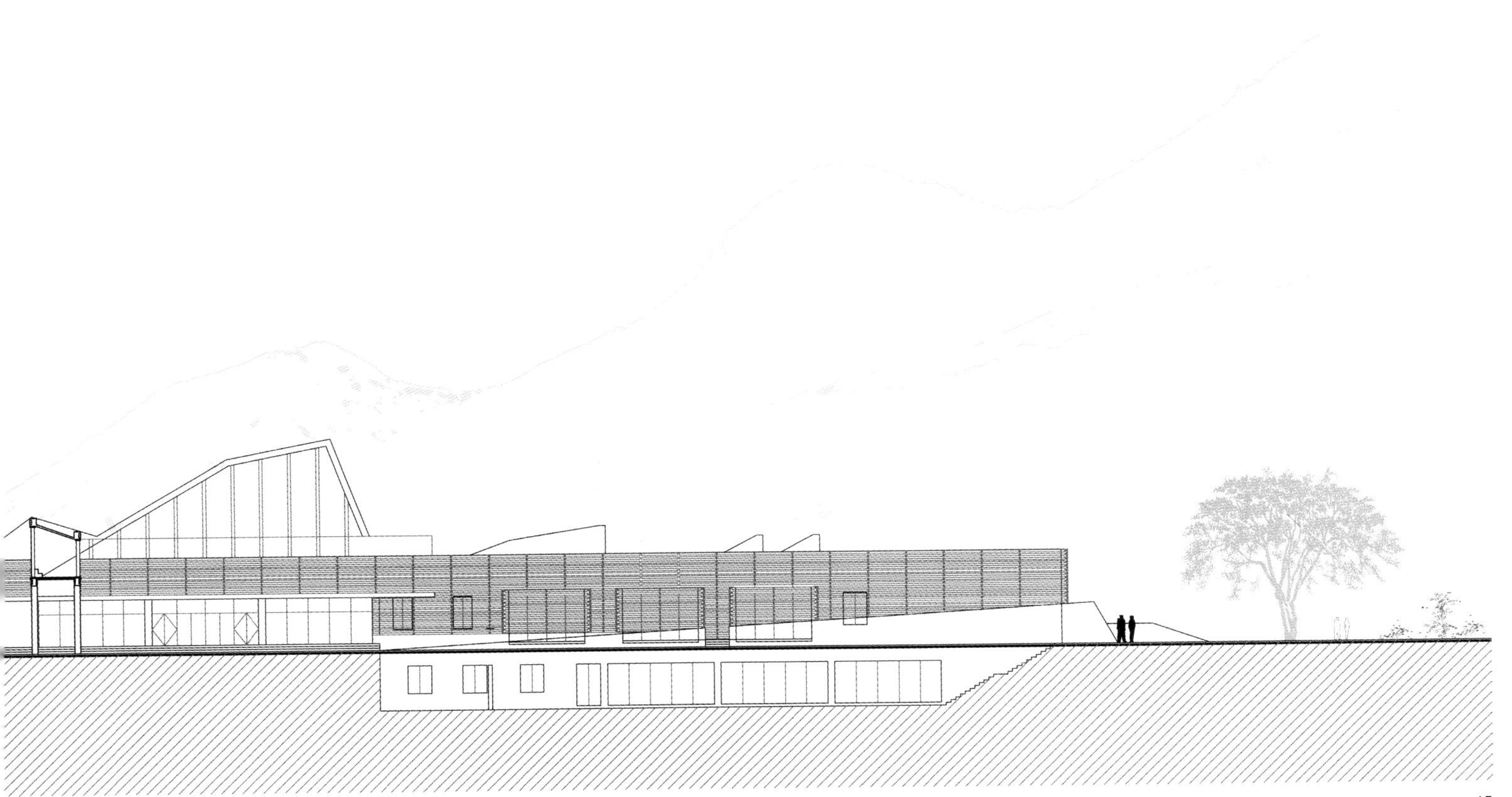

15

辉煌时代大厦 北京

Splendid Time Building Beijing

设计时间 / Design：2002

建成时间 / Completion：2004

建筑面积 / Building area：$64290m^2$

项目组 / Design team：崔彤 桂喆 平海峰 华夫荣 邝鸿军 陈长安

合作 / Collaborators：首钢设计研究院

业主 / Client：辉煌集团

摄影 / Photographer：杨超英 方振宁

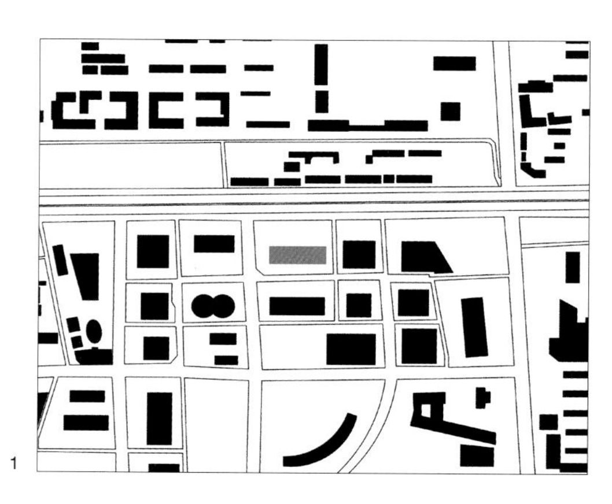

1 区位图
Location

2 建筑外景 1
Exterior 1

辉煌时代大厦坐落于北京中关村西区北部中段，紧邻北四环，与北京大学隔路相望。

设计过程中系统地思考基线、肌理、界面等秩序化元素，通过对位、对称、渐变等一系列设计手段，重新建立建筑与城市、建筑与场所、建筑与建筑之间的一种平衡体系，最终以求解的方式推导生成的新建筑，在必然与偶然、理性与浪漫、技术与艺术、简单与复杂之间寻找平衡的支点。

左右之间：设计手法是从左右相邻建筑中提取若干“灰色”属性要素，转化为具有自身平衡体系的一种语言与周边对话。作为中间角色，努力成为合唱队最认真的一员。

内外之间：顶部轮廓的内凹，是合理限高下主动地妥协，以便获得内外沟通而产生的视线通廊，并最大程度减少四环路中的阴影，在强调为城市作出贡献的同时，带给北部花园活力与生机。南北两个边庭是对两种外部空间做出不同尺度反应的共享大厅，边缘效应反映在外部空间被嵌入到建筑内部而形成的“城市边庭”。

上下之间：设计是基于界面关系、对位关系、基线网络思考而形成的外表平滑简洁、内部变化丰富的建筑形态。这种自下而上的渐变是在功能决定论下对“规矩”的重新思考。双角锥形式的逻辑源于分别设置于尽端的两个交通核心有节奏地收分所形成的金字塔形态，并且由表及里地体现出来，而且上下之间阶梯式空中花园形态的逻辑背后是每层办公空间具有三个方向的景观资源。

辉煌大厦自身的平衡体系，表现在隐约可见的对称图形下相互错动的三角锥体而构成的动态平衡体系。

The Splendid Time Building is located in the middle of the north side of the West District in Zhongguancun. It is close to the North Fourth Ring Road, facing Peking University across the road.

The design of the building takes into full consideration such elements as lines, textures and interfaces. By means of contrasting positions, symmetry and gradual changes, it keeps a balance between the building and the city, the building and the site, as well as the building and the surrounding buildings, which finally creates a new building. The building is a perfect combination of necessity and contingency, rationality and romanticism, technique and art, and simplicity and complexity.

Between the left and right：the ″grey color″ property is borrowed from the architecture on the left and right sides, and is made into a self-balanced color that corresponds to the surrounding environment. This building, standing among building groups, keeps perfect harmony with the other architecture in the area.

Between the inside and outside：an indentation is designed on the top of the building as a response to the height limit. This not only provides a good channel for integrating the outside and inside of the building, but also minimizes the shadow it might produce on the Fourth Ring Road. Therefore the design is beneficial for the city, and at the same time, brings vitality and vigor to the north garden of the city. However, the northern side hall and the southern side hall are designed in different ways to envelop the different exterior spaces to the north and south of the building. The exterior spaces are added to the inside of the building, forming an extraordinary edge effect and becoming the city's ″side halls″ .

Between the upper and lower：the design, taking into consideration the interfacial relationship, contrasting position relationship and baseline networks, forms not only an architecture with simple and smooth surfaces but also a dynamic and changeable interior structure. The building gradually changes from the bottom upwards, which is a challenge to ″function determinism″ . Two pyramids are designed at the two centers of the building to create a hanging garden between the upper and lower spaces so that people in the office spaces on each floor can enjoy the beautiful floral sight from three directions.

The Splendid Time Building is made up of a dynamic balance system constituted by the interactive cones under the looming symmetrical patterns.

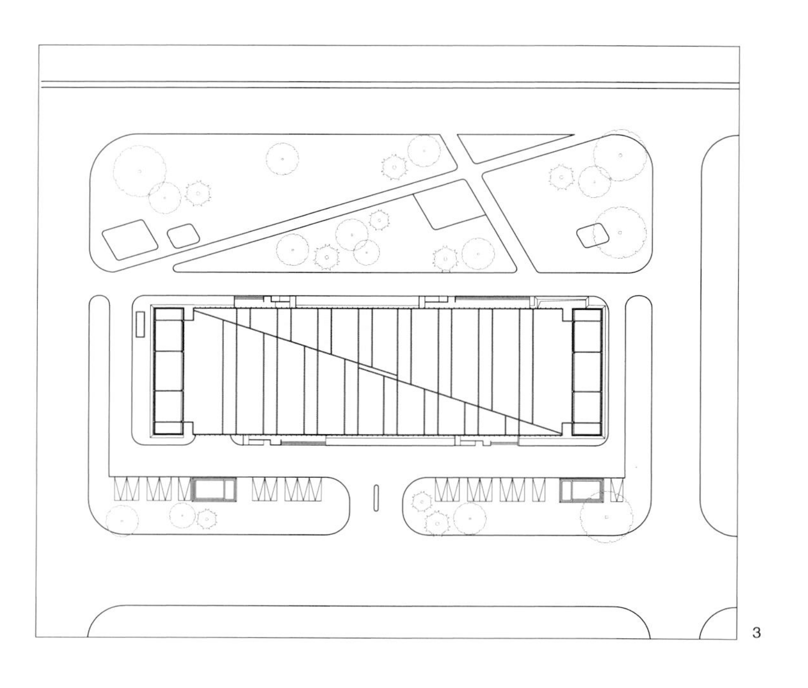

3

3 总平面图
Site Plan

4 建筑解析
Anlaysis

5 建筑外景 2
Exterior 2

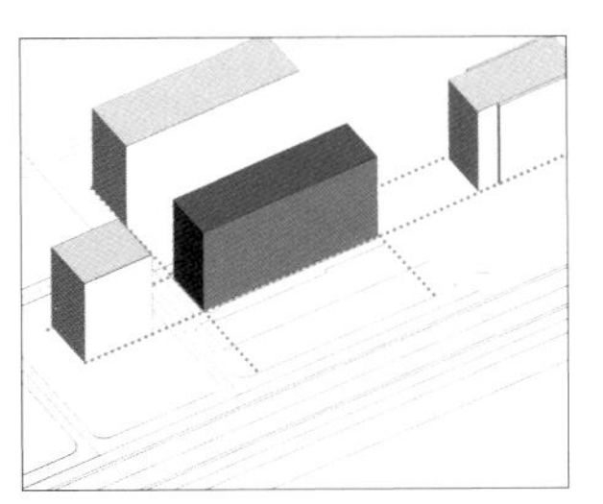

守边、对位
南北布局
Keep the edge, Counterpoint
North-South Layout

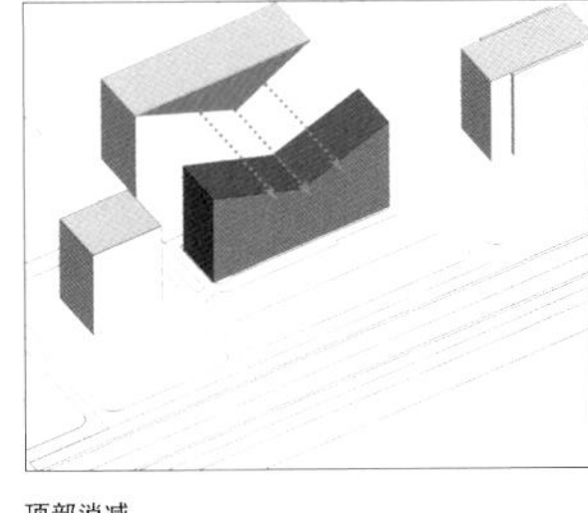

顶部消减
临楼视线通透
Top subduction
Maintain the view for next building

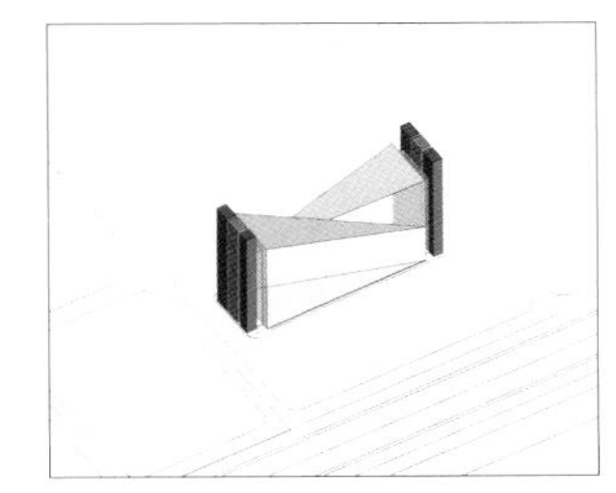

交通盒外置
自由划分平面
External traffic box
Free internal space division

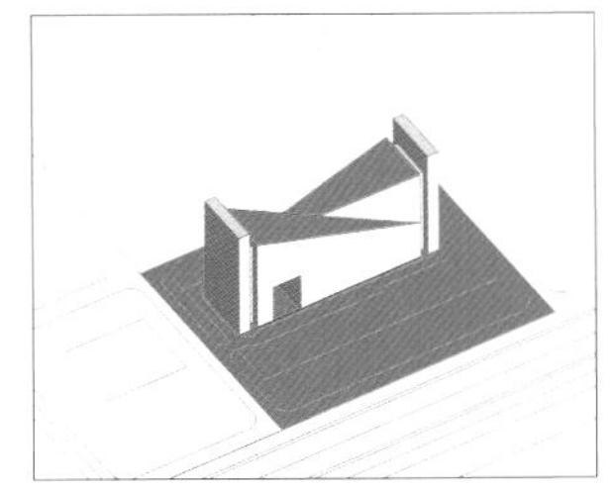

南北边框
几何求解
North South border
Geometric constraint solving

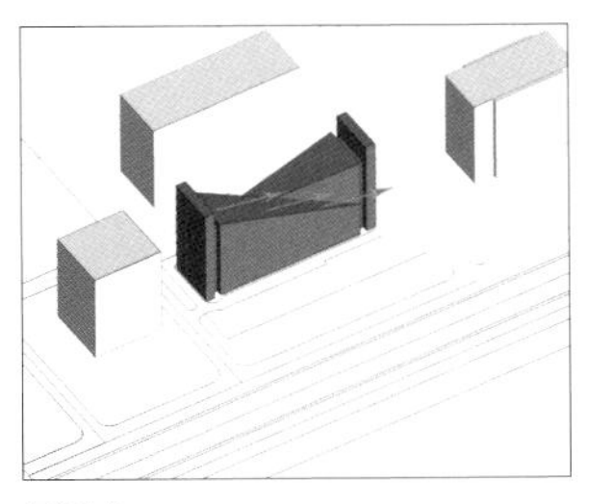

视线分析
西向景观最佳
View point analysis
Southwest is the best

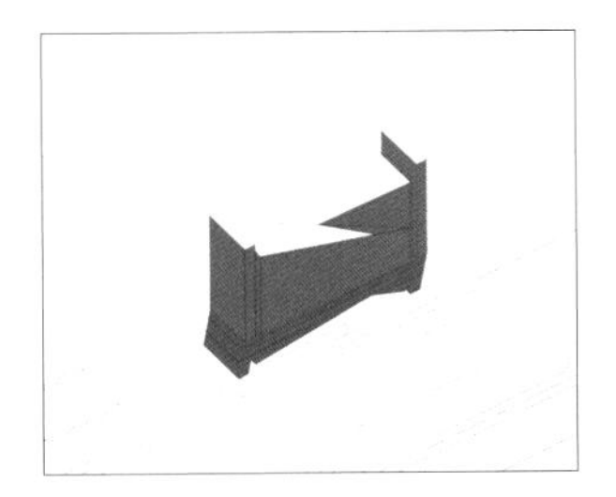

中部下凹
阴影减少
Center concave
Reduce shadow

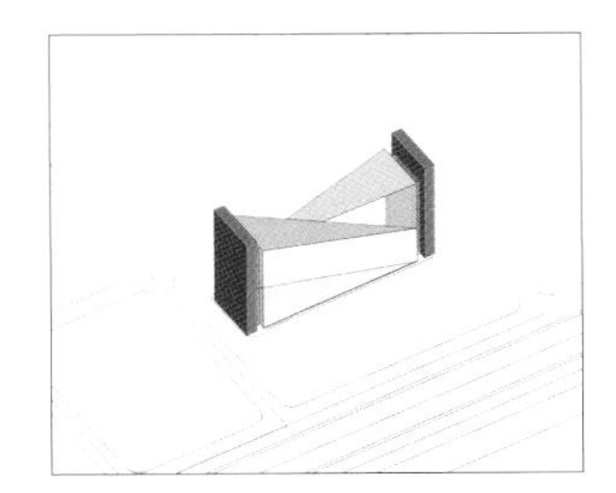

东西服务
南北采光通风
East-West Soild zone
North-South transparent zone

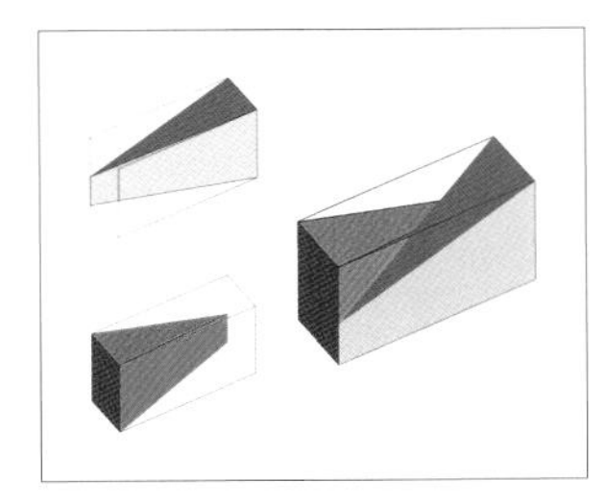

三角分析
矩形空间
Triangle analysis
Rectangular space

4

5

6

6 建筑外景 3
Exterior 3

7 首层平面
The 1st floor plan

8 二层平面
The 2nd floor plan

9 三层平面
The 3rd floor plan

10 七层平面
The 7th floor plan

11 八层平面
The 8th floor plan

12 九层平面
The 9th floor plan

13 十层平面
The 10th floor plan

14 十一层平面
The 11th floor plan

15 十二层平面
The 12th floor plan

16 十三层平面
The 13th floor plan

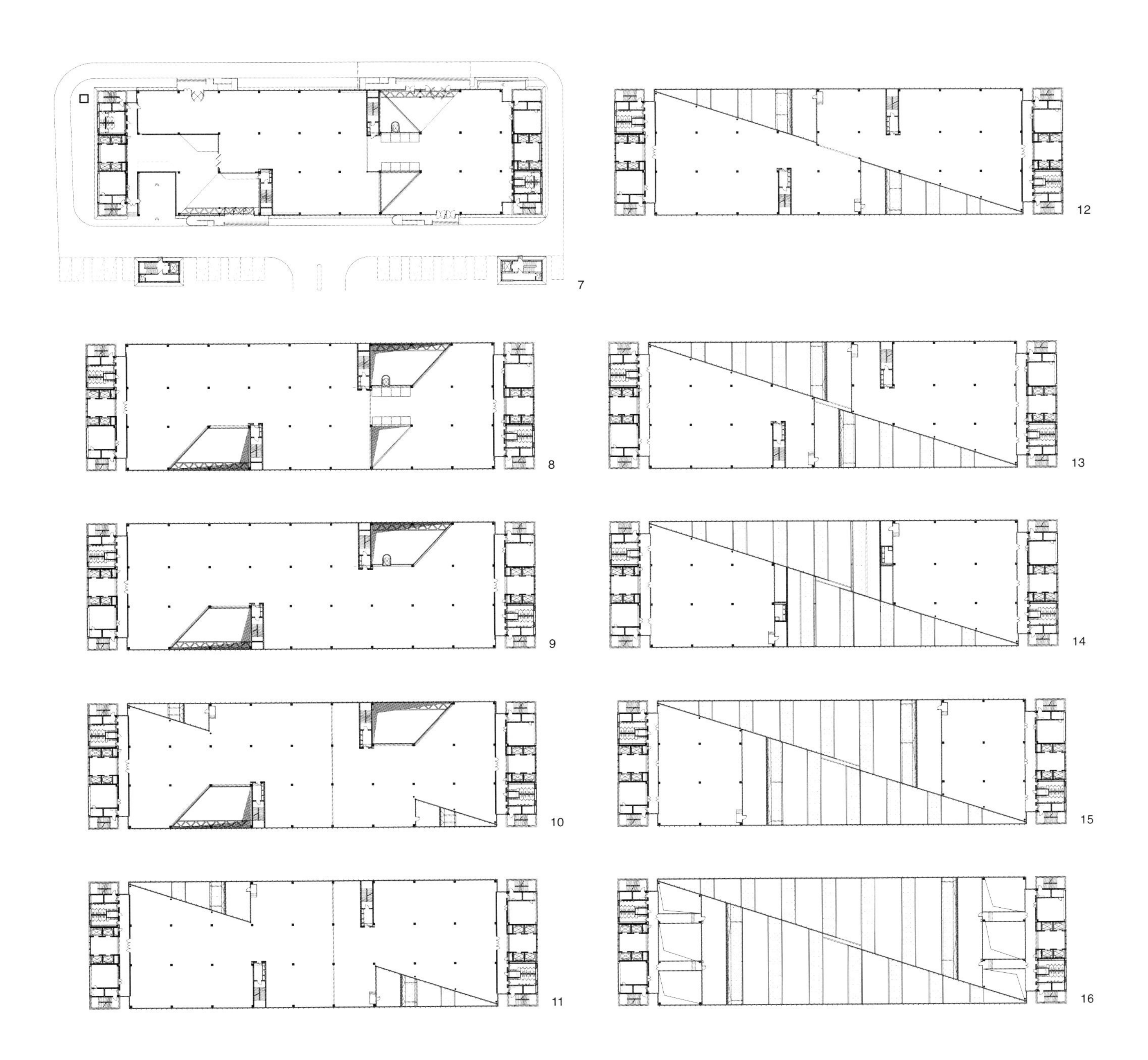

7
8
9
10
11
12
13
14
15
16

17 建筑外景 4
Exterior 4

18 建筑外景 5
Exterior 5

19 建筑室内 1
Interior 1

20 建筑室内 2
Interior 2

21 建筑外景 6
Exterior 6

22 建筑外景 7
Exterior 7

23 建筑外景 8
Exterior 8

17	19
18	20

21
22 | 23

歌华大厦 北京

Gehua Building Beijing

1 区位图
Location

2 建筑外景 1
Exterior 1

设计时间 / Design：2004
建成时间 / Completion：2006
建筑面积 / Building area：108500m^2
项目组 / Design team：崔彤 何川 黄文龙
业主 / Client：歌华集团
摄影 / Photographer：张广源 杨超英

歌华大厦的用途是文化产业创意基地。文化创意的特殊性区别于普通匀质空间的单一性而产生差异，差异并非由装修获得，它应由空间的方位、形态、尺度及组合方式决定。针对创意人员的行为研究，重新构筑了多中心组团。空间的重塑手法有三种：加法、减法和乘法。其一，将原来核心部分的大堂变为2层南北纵深的空间，减弱宏大尺度，增加使用效率，强化与街区和城市的联系，形成雍和宫地区创意产业园的开端式空间；其二，顶部空间向右延伸，加建2层高展览空间，构成"回望"旧城的主题场所，这一"动作"依赖于结构的支持，并提前将结构引起的系统变化作充分的考量；其三，作为重点地段的歌华大厦，它的地段优势不仅在于与旧城接壤和雍和宫毗邻，而且正好处于原护城河和旧城墙的位置。这一极度敏感的特质带我们回到梁思成先生的理想之中（"城墙上面积宽敞，可布置花池，栽种花草，安设公园椅，每隔若干距离的平台上可见凉亭供人游息，城墙和城楼上俯视护城河与郊外平原，远望西山远景或紫禁城宫殿，它将是世界上最特殊的公园之一，一个全长39.75km的立体环城公园"）。我们在屋顶上试图再造一种历史记忆。

形式依然是业主关注的焦点，最终形成的方案是在"城墙"、"青铜器"、"殿堂"三个方案基础上优化的结果，"青铜器"盒子基于歌华大厦作为创意基地而具有的独特品质。它以"G-box"雕刻了一个信息时代的建筑时空，借助典型的中国图案和文字的有机结合，试图恢复一个"技艺"和"记忆"，努力再现青铜时代的辉煌。而"回归"并非是伤感地回到某种历史废墟的吟诵，相反以还原的态度致力于空间和容器相关性研究，在建造和制造之间寻找精湛技艺的统一性。

The Gehua Building serves as a base for the cultural creation industry. The special features of cultural creation contribute to the diversity of the building that is distinct from the simplicity of the general space. Rather than being a result of design, the diversity was decided by means of its spatial orientation, shapes, dimensions and combinations. A multi-center complex was designed based on studies related to the habits of creative professionals. There are three ways to rebuild the space：addition, reduction and multiplication. Firstly, the lobby, which was formerly the core area of the building, was made into a two-story space running south to north. This not only helped reduce the grand size and increase efficient use of the space, but also strengthened its connection with the streets and city, to form an open space of the creative industrial park near the Lama Temple. Secondly, the headspace was extended to the right, and a two-story exhibition area added, forming a theme area to "reflect" on the old city. This is done by relying on the structure of the building with the full consideration of any possible system changes of the structure. Thirdly, located at an important area of Beijing, the Gehua Building has many advantages demonstrated not only in its adjacency to the old parts of Beijing and the Lama Temple, but also its closeness to the original city moat and ancient city wall. This unique location reminds us of Mr. Liang Sicheng's ideal design："the city wall is spacious. Flowers, pools, grasses and benches are arranged inside it. Pavilions for people to take a rest are designed at the lookout towers at regular intervals. Standing on the city wall and the city gate tower, people can see the moat, the outside view of the city, the Western Hills far away, and the Forbidden City inside the wall. It will become one of the most unique parks in the world, a three-dimensional park in a full length of 39.75km which stretches around the whole city." That description is our attempt to restore a historical link on the roof of the Gehua Building.

2

3

4

The building's shape was a major concern of the owner. Bronze ware, which has unique characteristics suited for a cultural creation base, was an optimum choice based on three options："city wall"，"bronze ware" and "palace"．By combining traditional Chinese patterns with Chinese characters, the design tries to restore artistry and recall memories of the Bronze Age and reproduce such glorious time in ancient Chinese art. However, a retrospective of old times does not mean mourning the historical ruins in the past, but rather to find a relationship between the space and vessels, and to discover a unity between building construction and vessel creation.

3 街景展示
Streetscape

4 模型
Model

5 总平面图
Site plan

6 建筑外景 2
Exterior 2

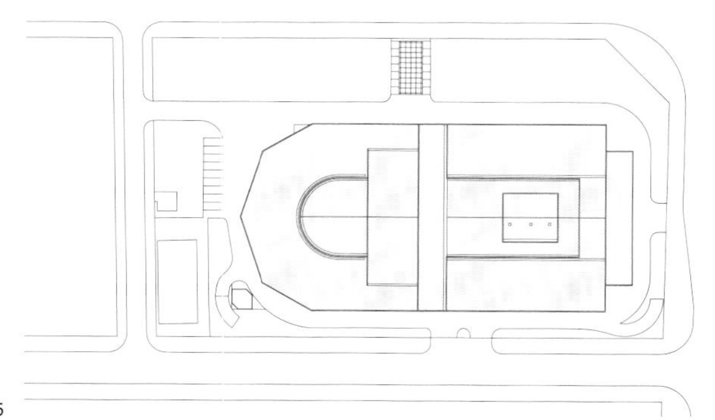

5

6

8

9

7 建筑外景 3
Exterior 3

8 建筑外景 4
Exterior 4

9 建筑外景 5
Exterior 5

光大国际中心 北京

Everbright World Center Beijing

设计时间 / Design：2004

建成时间 / Completion：2007

建筑面积 / Building area：120000m^2

项目组 / Design team：崔彤 王欣 刘向志 文业清

业主 / Client：光大国际中心

摄影 / Photographer：杨超英

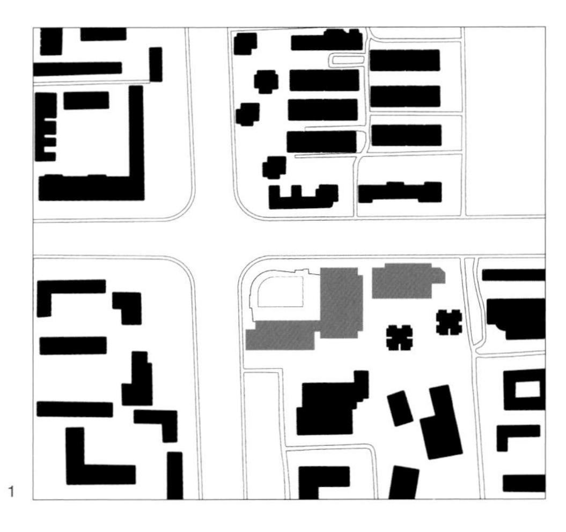

1 区位图
Location

2 建筑外景 1
Exterior 1

建筑设计并不是主观性的无限发挥，也不是偶发性的某种巧合，它是解决问题的一个过程。设计的结果是面对场所特征、使用功能、商业动态提出解决问题的答案。在这个解决矛盾的过程中，伴随着客观条件限制下的必然性寻求。光大国际中心的建筑形态融合于简洁手法和复杂关系的统一体中，以一种还原的态度，执着于传统文化信息的表达，用〝记忆〞的建筑语言，重新建立一种新秩序。

北京城的几何对称性导致城市意象的对称性，以中轴线为对称轴的城市认知图像，如东四、西四；东单、西单；东直门、西直门的客观存在。光大国际中心作为北京城中重要的场所，试图通过城市认知图像的对称性恢复京城应有的规律。

光大国际中心作为城市节点的重要建筑，落在两条道路的〝开端〞和〝终结〞处，并聚焦十字路口。它的这种双重性一方面体现在平安大街的传统文脉；另一方面体现在金融街国际化的水准。这种双重气质的融贯成为此时此地的地标建筑。

源于场所分析和形态解析，光大国际中心的独特性有别于塔式高层建筑的表现而形成卓尔不凡的品质。坚持批判〝表皮主义〞的同时，为如此体量、身高的建筑寻求一种修正式的表皮，在综合考虑尺度、边界、对位的基础上，希望超越所谓建筑的遮蔽体而成为有意义的界面。

整体性体现在三栋高层建筑与京剧院在功能上的独立和形态上的有机统一，而达到和而不同的境界。它不仅表现在单体形象完整，群体形象完美，更主要的是相互因借而形成的〝场力〞，这种新秩序，首先存在于面对城市所形成的完整界面关系和外向性仪典格局中，同时体现在三者对话关系中，〝背景〞对〝角色〞的互补效应。

通过一种引人思索的抽象手法去探究隐含在中国传统表面之后的〝存在〞，应用律动变化的图案组合希望记录和谱写更多的恒久，并以怀旧式态度，借用博古架的结构秩序承载几分厚重。

Architectural design is not an infinite elaboration of subjectivity, neither an accidental coincidence, but is rather a process of solving problems. The result of a design is to give solutions to a building's site features, function, and commercial trends. Such process is also a logical pursuit under objective conditions. The building form of Everbright World Center presents both simplicity in design and intricacy in relationship. It focuses on presenting traditional culture in design and building a new style by recalling architectural features through memories.

The symmetrical image of Beijing is a result of the geometric symmetrical features of the city. The Chinese capital's distinct images, such as Dongsi and Xisi, Dongdan and Xidan, Dongzhimen and Xizhimen, are symmetrical to the central axis of Beijing. Everbright World Center, as a significant venue in the city, tries to restore and continue the proper pattern by putting an emphasis on the symmetrical image of Beijing.

Everbright World Center is a distinguished-looking building in the city. It is located at the beginning and end of two roads, and is facing an intersection. Its duality, on the one hand, inherits the traditional features of Ping'an Street, and on the other, shows the international level of Financial Street. Such duplicity makes it a unique landmark building in this area in the 21st century.

The uniqueness of Everbright World Center is different from skyscrapers; it has its own outstanding qualities and features. The building tries to create a new style of surface for a building with such massive volume and height while contrasting against the 〝surface phenomenon〞 of modern buildings. By taking dimension, boundary and location into full consideration, the design makes this building into a piece of meaningful art rather than simply a place of shelter.

The center is an integrated one because its three high-rise buildings that create a harmonious and distinct atmosphere in its independent functions and uniform shapes, quite different from the Beijing Opera Museum facing it. Everbright World Center has a complete image for single buildings and a complex as a whole. More important, the two images serve as a foil to each other.

The design explores the cultural existence hiding behind Chinese traditions by using an abstract artistic method. It tries to create an everlasting piece of art by dynamics and changing pattern combinations, and it inherits the rich traditional culture by means of the 〝curio shelf〞 surface style.

2

3

4

5

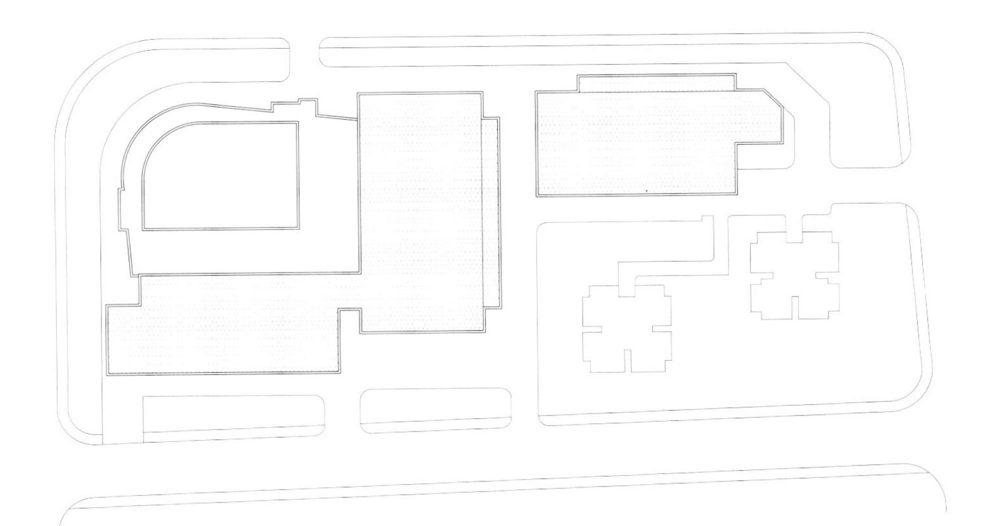

6

3 手稿
Sketch

4 建筑外景 2
Exterior 2

5 建筑外景 3
Exterior 3

6 总平面图
Site plan

7

7 建筑外景 4
Exterior 4

8 模型
Model

9 建筑细部
Exterior details

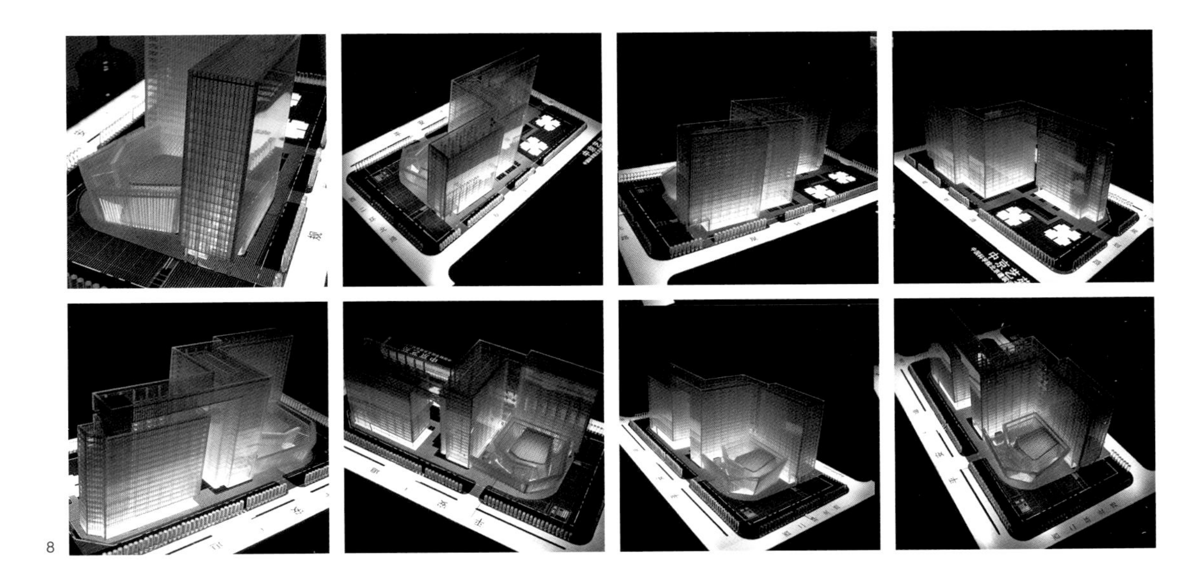

8

9

作品年表

Chronology of Works

空间之间的建构

中国科学院国家科学图书馆
National Science Library of the CAS

地点 / Location：北京 / Beijing
设计 / Design：1999
建成 / Completion：2002
建筑面积 / Building area：41000m^2
项目组 / Team：崔彤 白小菁 夏炜 王知非 宋爽 邝红军
业主 / Client：中科院文献情报中心
摄影 / Photographer：杨超英 傅兴

获国家第十一届优秀工程设计金奖（2004 年）
获建设部优秀勘察设计一等奖（2004 年）
获北京市第十一届优秀工程设计一等奖（2003 年）
获全国第十一届优秀工程设计金质奖（2004 年）
获中国建筑学会建筑创作大奖（2009 年）
获第四届中国建筑学会建筑创作优秀奖（2006 年）
获建设部部级城乡建设优秀勘察设计一等奖（2004 年）
获第六届首都建筑设计汇报展建筑创作二等奖及十佳建筑奖
获国家鲁班奖

国家开发银行（复内 4-2 项目）
China Development Bank(Porject 4-2)

地点 / Location：北京 / Beijing
设计 / Design：2006
建成 / Completion：2013
建筑面积 / Building area：151000m^2
项目组 / Team：崔彤 赵正雄 何川 潘华 王欣 桂喆
业主 / Client：国家开发银行
摄影 / Photographer：杨超英

获第十三届首都城市规划建筑设计方案汇报展公共建筑优秀设计方案二等奖（2006 年）
获北京市第十八届优秀工程设计一等奖（2015 年）
获全国优秀工程勘察设计行业奖公共建筑一等奖（2015 年）

泰国曼谷·中国文化中心
The Chinese Cultural Center in Bankok

地点 / Location：泰国 曼谷 / Bankok, Thailand
设计 / Design：2008
建成 / Completion：2012
建筑面积 / Building area：7656m^2
项目组 / Team：崔彤 王一钧 桂喆 吕僖 陈希 周军 房木生 苑蕾
业主 / Client：中国文化部
摄影 / Photographer：刘崇明 范虹

获首都第十八届城市规划建筑设计方案汇报展优秀方案奖（2011 年）
获北京市第十八届优秀工程设计一等奖（2015 年）
获全国优秀工程勘察设计行业奖建筑工程公共建筑一等奖（2015 年）

法国巴黎·中国文化中心
The Chinese Cultural Center in Paris

地点 / Location：法国 巴黎 / Paris, France
设计 / Design：2002
建成 / Completion：2006
建筑面积 / Building area：4100m²
项目组 / Team：崔彤 赵霁欣 颜胜男
业主 / Client：中国文化部

中国工艺美术馆
National Arts and Crafts Museum of China

地点 / Location：北京 / Beijing
设计 / Design：2012
建成 / Completion：N/ A
建筑面积 / Building area：86800m²
项目组 / Team：崔彤 刘建平 赵迎 王一钧 张润欣 王康 陈希 司亚琨 曹炜 房木生 周军 许楠
业主 / Client：中国文化部、中国艺术研究院

中国驻贝宁大使馆
The Chinese Embassy in Benin

地点 / Location：贝宁 / Benin
设计 / Design：2006
建成 / Completion：建设中 / under construction
建筑面积 / Building area：8442m²
项目组 / Team：崔彤 黄文龙 王欣
业主 / Client：中国外交部

国家图书馆二期暨国家数字图书馆
Second Phase of National Library - National Digital Library

地点 / Location：北京 / Beijing
设计 / Design：1999
建成 / Completion：N/ A
建筑面积 / Building area：79080m²
项目组 / Team：崔彤 王欣 桂喆
合作者 /Co：德国 ABB
业主 / Client：国家图书馆

中国驻巴西圣保罗总领事馆
Chinese Consulate-General in Sao Paulo, Brazil

地点 / Location：巴西 圣保罗 / Sao Paulo, Brazil
设计 / Design：2006
建成 / Completion：2013
建筑面积 / Building area：4645m^2
项目组 / Team：崔彤 何川 赵正雄 罗大坤 韩勇玮
业主 / Client：中国外交部

中国国家美术馆
National Arts Museum of China

地点 / Location：北京 / Beijing
设计 / Design：2006
建成 / Completion：N/ A
建筑面积 / Building area：39673m^2
项目组 / Team：崔彤 王欣 何川 白小菁 桂喆
业主 / Client：中国美术馆

获国际方案设计竞赛二等奖

中国驻埃塞俄比亚大使馆
Chinese Embassy in Ethiopia

地点 / Location：埃塞俄比亚 亚地斯亚巴贝 / Addis Ababa, Ethiopia
设计 / Design：2006
建成 / Completion：2013
建筑面积 / Building area：7966m^2
项目组 / Team：崔彤 黄文龙 韩勇炜 于义夫 张润欣
业主 / Client：中国外交部

国家动物博物馆
National Zoological Museum of China

地点 / Location：北京 / Beijing
设计 / Design：2002
建成 / Completion：2006
建筑面积 / Building area：42900m^2
项目组 / Team：崔彤 桂喆 何川 白小菁 文业清
业主 / Client：中国科学院动物研究所
摄影 / Photographer：舒赫 杨超英

获全国优秀工程勘探设计银奖（2010 年）
获北京市第十四届优秀工程设计一等奖（2009 年）
第六届中国建筑学会建筑创作佳作奖（2011 年）
获全国优秀工程勘察设计建筑工程二等奖（2009 年）

中国民政部部委办公楼
The Office Building of Chinese Civil Affair Ministry

地点 / Location：北京 / Beijing
设计 / Design：2007
建成 / Completion：N/ A
建筑面积 / Building area：51450m²
项目组 / Team：崔彤 赵正雄 何川 苏东坡 韩勇炜
业主 / Client：中国民政部

中国教育部信息化大厦
Informationization Building of Chinese Education Ministry

地点 / Location：北京 / Beijing
设计 / Design：2011
建成 / Completion：N/ A
建筑面积 / Building area：106000m²
项目组 / Team：崔彤 张润欣 刘建平 赵迎 王一钧 唐璐 陈希
业主 / Client：中国教育部

中国人民银行总行新办公楼
New Head Office of The People's Bank of China

地点 / Location：北京 / Beijing
设计 / Design：2011
建成 / Completion：N/ A
建筑面积 / Building area：173500m²
项目组 / Team：崔彤 刘建平 赵迎 张润欣 王一钧 唐璐 陈希
业主 / Client：中国人民银行

中国科学院化学所前沿交叉研究平台 B 楼
Macromolecule Research Building B of Chemical Institute, CAS

地点 / Location：北京 / Beijing
设计 / Design：2008
建成 / Completion：2012
建筑面积 / Building area：23700m²
项目组 / Team：崔彤 何川 张润欣 黄文龙 刘立森 苏东坡 华夫荣 刘建平
业主 / Client：中国科学院化学研究所
摄影 / Photographer：杨超英

北京林业大学学研中心
Acadmic Research Center of Bejing Forestry Uinversity

地点 / Location：北京 / Beijing
设计 / Design：2008
建成 / Completion：建设中 / under construction
建筑面积 / Building area：96000m²
项目组 / Team：崔彤 何川 罗大坤 苏东坡 张润欣 华夫荣 桂喆 唐璐 池依娜 辛鑫
业主 / Client：北京林业大学
摄影 / Photographer：杨超英 范虹

获首都第十八届城市规划建树设计汇展公共建筑优秀设计方案三等奖
获北京市第十八届优秀工程设计二等奖（2015 年）

北川央企办公楼群
Central Enterprise Office Buildings of Beichuan

地点 / Location：北川 / Beichuan
设计 / Design：2009
建成 / Completion：2011
建筑面积 / Building area：50000m²
项目组 / Team：崔彤 赵正雄 张润欣 辛鑫 唐璐 韩勇炜
业主 / Client：北川县人民政府

中国科学院计算技术研究所
Institute of Computing Technology, CAS

地点 / Location：北京 / Beijing
设计 / Design：2002
建成 / Completion：2004
建筑面积 / Building area：30000m²
项目组 / Team：崔彤 王欣 李昕滨 文业清
业主 / Client：中国科学院计算技术研究所
摄影 / Photographer：杨超英 舒赫

获全国优秀工程勘探设计建筑工程三等奖（2008 年）
获建筑部优秀工程建筑设计二等奖（2007 年）

中国科学院电子研究所科研综合楼
Scientific Research Building of Institute of Electronics, CAS

地点 / Location：北京 / Beijing 海淀区中关村
设计 / Design：2003
建成 / Completion：2005
建筑面积 / Building area：24248m²
项目组 / Team：崔彤 赵正雄 曾荣 时胜荣
业主 / Client：中国科学院电子学研究所
摄影 / Photographer：杨超英

中国科学院研究生院教学楼（中关村园区）
Teaching Building of the Graduate University of CAS, in Zhong guan cun Park

地点 / Location：北京 / Beijing
设计 / Design：2001
建成 / Completion：2003
建筑面积 / Building area：65300m²
项目组 / Team：崔彤 夏炜 王知非 王毓琳
业主 / Client：中国科学院研究生院
摄影 / Photographer：傅兴 杨超英

获建设部部级优秀勘探设计三等奖（2006 年）
获北京市第十二届优秀工程设计一等奖（2005 年）
获第八届首都建筑设计汇报公共建筑设计方案二等奖（2002 年）
获第八届首都建筑设计汇报公共建筑十佳设计方案奖（2002 年）

中国科学院数学与系统科学研究院科研楼
Scientific Research Building of Acadmey of Mathematics and Systerms Science, CAS

地点 / Location：北京 / Beijing
设计 / Design：2002
建成 / Completion：2006
建筑面积 / Building area：42900m²
项目组 / Team：崔彤 何川 白小菁 王一钧 桂喆 韩勇炜 张润欣
业主 / Client：中国科学院数学研究院
摄影 / Photographer：杨超英

东莞理工学院教学楼
The Teaching Building of Dongguan University of Technology

地点 / Location：东莞 / Dongguan
设计 / Design：2002
建成 / Completion：2004
建筑面积 / Building area：24000m²
项目组 / Team：崔彤 许明菁 赵霁欣 鲁岩
业主 / Client：东莞理工学院
摄影 / Photographer：杨超英

获东莞市优秀建筑设计一等奖（2005 年）

中国科学院动物研究所
Institute of Zoology, CAS

地点 / Location：北京 / Beijing
设计 / Design：2002
建成 / Completion：2006
建筑面积 / Building area：42900m²
项目组 / Team：崔彤 桂喆 何川 白小菁
业主 / Client：中国科学院动物研究所
摄影 / Photographer：杨超英 舒赫

获全国优秀工程勘探设计银奖（2010 年）
获北京市第十四届优秀工程设计一等奖（2009 年）
第六届中国建筑学会建筑创作佳作奖（2011 年）
获全国优秀工程勘察设计建筑工程二等奖（2009 年）

721 国家重点工程
721 National Key Project

地点 / Location：北京 / Beijing
设计 / Design：2006
建成 / Completion：2010
建筑面积 / Building area：52000m²
项目组 / Team：崔彤 刘向志 于鹏 桂喆 韩冰浩
业主 / Client：
摄影 / Photographer：杨超英

化学工业出版社
Chemical Industry Press

地点 / Location：北京 / Beijing
设计 / Design：2004
建成 / Completion：2006
建筑面积 / Building area：11000m²
项目组 / Team：崔彤 桂喆 何川 黄文龙
业主 / Client：化学工业出版社
摄影 / Photographer：杨超英

获第十一届首都城市规划建筑设计方案汇报展建筑节能设计三等奖（2005 年）
获公共建筑优秀设计方案三等奖（2005 年）

中国遥感卫星地面站总部
Headquarter Center for Earth Observation and Digital Earth, CAS

地点 / Location：北京 / Beijing
设计 / Design：2007
建成 / Completion：2010
建筑面积 / Building area：31800m²
项目组 / Team：崔彤 罗大坤 苏东坡 曾荣 何川 赵正雄
业主 / Client：中科院中国遥感卫星地面站
摄影 / Photographer：杨超英

获北京市第十七届优秀工程设计奖公共建筑二等奖
获 2013 年全国优秀工程勘查设计行业奖公共建筑三等奖

中国科学院研究生院新校区工程
New Campus of Graduate University of Chinese Acadmey of Science

地点 / Location：北京 / Beijing
设计 / Design：2008
建成 / Completion：建设中 / under construction
建筑面积 / Building area：310000m²
项目组 / Team：崔彤 赵正雄 白小菁 许明菁 罗大坤 黄文龙 桂喆
刘立森 王一钧 张润欣 苏东坡 吕僖 刘建平 刘颖
韩勇炜 江岩 宋小慧 华夫荣 顾国端 杨俊 辛鑫 唐璐
业主 / Client：中国科学院研究生院

获奖获北京市优秀城乡规划设计三等奖（2011 年）

中国农业大学烟台新校区工程
Yantai Campus of China Agricultral Unversity

地点 / Location：烟台 / Yantai
设计 / Design：2003
建成 / Completion：2005
建筑面积 / Building area：960000m²
项目组 / Team：崔彤 赵正雄 于鹏 于义夫 陆保新 连娜 袁朵 李洪刚
业主 / Client：中国农业大学

获年度优秀建筑设计项目一等奖（2005）

上海建平中美学校
Shanghai Jianping Experimental Middle School

地点 / Location：上海 / Shanghai
设计 / Design：1998
建成 / Completion：2000
建筑面积 / Building area：101600m²
项目组 / Team：崔彤 王知非 邹凡 李军
业主 / Client：上海建平中学

内蒙古科技大学逸夫楼
Yifu Building, Inner Mongolia University of Science and Technology

地点 / Location：内蒙古 / Inner Mongolia 包头市昆区
设计 / Design：2009
建成 / Completion：2012
建筑面积 / Building area：23000m²
项目组 / Team：崔彤 赵正雄 曾荣 唐璐
业主 / Client：内蒙古科技大学

北京石油化工学院图书馆
Library of Beijing Institute of Petrochemical Technology

地点 / Location：北京 / Beijing
设计 / Design：2011
建成 / Completion：建设中 / under construction
建筑面积 / Building area：21000m²
项目组 / Team：崔彤 白小菁 郭亚男 辛鑫
业主 / Client：北京石油化工学院

东北大学新校区宿舍（浑南校区）
Dormitory in Hunnan Campus of Dongbei University

地点 / Location：沈阳 / Shenyang
设计 / Design：2012
建成 / Completion：建设中 / under construction
建筑面积 / Building area：78830m²
项目组 / Team：崔彤 王一均 刘建平 曹炜 杨皞
业主 / Client：东北大学

大唐国际煤化铝国家重点实验室
State Key Laboratory of Coalification Aluminum, China Datang Corporation

地点 / Location：内蒙古鄂尔多斯
设计 / Design：2012
建成 / Completion：建设中 / under construction
建筑面积 / Building area：15600m²
项目组 / Team：崔彤 苏东坡 赵正雄
业主 / Client：中国大唐国际集团

中科院上海同步辐射中心
Shanghai Synchrotron Radiation Facility, Chinese Academy of Sciences

地点 / Location：上海 / Shanghai
设计 / Design：1999
建成 / Completion：N/ A
建筑面积 / Building area：61500m²
项目组 / Team：崔彤 白小菁 黄文龙 梅咏 王知非
业主 / Client：中国科学院

获全国设计竞赛一等奖

中国科学院生态环境研究中心生态楼
Ecological Building of Research Center for Eco-Environmental Sciences, Chinese Academy of Sciences

地点 / Location：北京 / Beijing
设计 / Design：2001
建成 / Completion：2002
建筑面积 / Building area：8100m²
项目组 / Team：崔彤 刘向志 文业清
业主 / Client：中国科学院生态环境研究中心
摄影 / Photographer：杨超英

获北京市第十一届优秀工程建筑设计二等奖（2003 年）

中科院科技创新与信息平台
Headquarter Office Building of CAS
Science & Technology Innovation & Information Platform

地点 / Location：北京 / Beijing
设计 / Design：2008
建成 / Completion：建设中 / under construction
建筑面积 / Building area：32000m²
项目组 / Team：崔彤 何川 王康 刘建平 桂喆 赵迎
业主 / Client：中国科学院

中国科学院钱学森科学实验室
Qian Xuesen Science Laboratory, CAS

地点 / Location：北京 / Beijing
设计 / Design：2008
建成 / Completion：2012
建筑面积 / Building area：55000m²
项目组 / Team：崔彤 桂喆 韩勇炜 刘建平 辛鑫
业主 / Client：中国科学院

光大国际中心
Everbright World Center

地点 / Location：北京 / Beijing
设计 / Design：2004
建成 / Completion：2007
建筑面积 / Building area：120000m²
项目组 / Team：崔彤 王欣 刘向志 文业清
业主 / Client：光大国际中心
摄影 / Photographer：杨超荣

歌华大厦
Gehua Building

地点 / Location：北京 / Beijing
设计 / Design：2004
建成 / Completion：2006
建筑面积 / Building area：108500m²
项目组 / Team：崔彤 何川 黄文龙
业主 / Client：北京歌华集团
摄影 / Photographer：张广源 杨超英

天津滨海新区于家堡起步区新金融办公楼
New Finance Office Building in Tianjing Yujiapu Financial District

地点 / Location：天津 / Tianjing
设计 / Design：2008
建成 / Completion：2012
建筑面积 / Building area：90390m^2
项目组 / Team：崔彤 赵正雄 白小菁 罗大坤 刘立森 唐璐 刘颖
业主 / Client：

辉煌时代大厦
Splendid Time Building

地点 / Location：北京 / Beijing
设计 / Design：2002
建成 / Completion：2004
建筑面积 / Building area：64290m^2
项目组 / Team：崔彤 赵嘉康 桂喆 平海峰 华夫荣 邝鸿钧 陈长安
业主 / Client：辉煌集团
摄影 / Photographer：杨超英 方振宁

获建设部2005年度部级城乡优秀勘察设计二等奖
获北京市第十二届优秀建筑设计二等奖（2005年）

京东商城总部办公楼
Headquarter Office Building of 360buy Jingdong Mall

地点 / Location：北京 / Beijing
设计 / Design：2011
建成 / Completion：建设中 / under construction
建筑面积 / Building area：283400m^2
项目组 / Team：崔彤 何川 刘立森 白小菁 刘建平 赵迎 郭亚男 苏东坡 王康 唐明 汪国顺
业主 / Client：北京京东世纪贸易有限公司

科实信息城
Keshi Information Tower

地点 / Location：北京 / Beijing
设计 / Design：2002
建成 / Completion：2004
建筑面积 / Building area：86500m^2
项目组 / Team：崔彤 白小菁 邝鸿钧 李军 穆斌 时胜荣 芮文宣 李哲
业主 / Client：

获第七届首都建筑设计汇报展建筑设计创作设计二等奖及十佳建筑奖

北京立思辰办公楼
Beijing Lanxum Office Building

地点 / Location：北京 / Beijing
设计 / Design：2011
建成 / Completion：建设中 / under construction
建筑面积 / Building area：63000m^2
项目组 / Team：崔彤 罗大坤 苏东坡 刘建平 赵迎 王康
业主 / Client：北京立思辰网络科技有限公司

物华科技综合楼
Wuhua Technology Integrity Building

地点 / Location：北京 / Beijing
设计 / Design：2007
建成 / Completion：N/ A
建筑面积 / Building area：46890m^2
项目组 / Team：崔彤 桂喆
业主 / Client：物华科技有限公司

河南省电力公司综合调度楼
Integrity Dispatching Building of Henan Electric Power Corporation

地点 / Location：郑州 / Zhenzhou
设计 / Design：2006
建成 / Completion：N/ A
建筑面积 / Building area：83000m^2
项目组 / Team：崔彤 桂喆 刘立森
业主 / Client：河南省电力公司

中国钢铁大厦
China Steel Tower

地点 / Location：北京 / Beijing
设计 / Design：2011
建成 / Completion：N/ A
建筑面积 / Building area：855000m^2
项目组 / Team：崔彤 桂喆 罗大坤 查季云
业主 / Client：首钢集团

东直门交通枢纽暨东华广场
Dongzhimen Transportation Hub - Donghua Square

地点 / Location：北京 / Beijing
设计 / Design：2004
建成 / Completion：N/ A
建筑面积 / Building area：830000m^2
项目组 / Team：崔彤 王欣 梅咏
业主 / Client：北京东华门房地产公司

获全国设计竞赛一等奖

郑州金印财富广场
Zhenzhou Golden Seal Fortune Plaza

地点 / Location：郑州 / Zhenzhou
设计 / Design：2004
建成 / Completion：2006
建筑面积 / Building area：127200m^2
项目组 / Team：崔彤 李昕滨 于鹏 于义夫 潘华
业主 / Client：河南安华房地产公司

获 2013 全国人居经典建筑规则设计方案竞赛建筑 / 环境金奖

山东济宁反恐中心
Jining Counterterrorism Center, Shandong Province

地点 / Location：济宁 / Jining
设计 / Design：2012
建成 / Completion：建设中 / under construction
建筑面积 / Building area：55000m^2
项目组 / Team：崔彤 何川 刘建平 赵迎 刁建存 苏东坡 王康
业主 / Client：山东济宁公安局

新奥集团总部
Enn Headquarters

地点 / Location：廊坊 / Langfan
设计 / Design：2008
建成 / Completion N/ A
建筑面积 / Building area：86500m^2
项目组 / Team：崔彤 黄文龙 潘华 韩勇炜
业主 / Client：新奥集团

北川抗震纪念园
Beichuan Earthquake Memorial Park

地点 / Location：北川 / Bei Chuan
设计 / Design：2009
建成 / Completion N/ A
建筑面积 / Building area：6000m²
项目组 / Team：崔彤 赵正雄 孟凡理 祁东阳
业主 / Client：四川北川县人民政府

吴为山雕塑馆工作室
Wu Weishan Sculpture Studio

地点 / Location：北京 / Beijing
设计 / Design：2011
建成 / Completion：建设中 / under construction
建筑面积 / Building area：5200m²
项目组 / Team：崔彤 刘建平 陈希 吕倩
业主 / Client：吴为山

北京天竺国展商业金融中心
Beijing Tianzhu International Exhibition & Finance Center

地点 / Location：北京 / Beijing
设计 / Design：2009
建成 / Completion N/ A
建筑面积 / Building area：115000m²
项目组 / Team：崔彤 罗大坤 苏东坡
业主 / Client：北京澳金园置业发展有限公司

玉树游客服务中心
Yushu Visitor Center

地点 / Location：青海玉树 / Qinghai Yushu
设计 / Design：2010
建成 / Completion：2013
建筑面积 / Building area：4500m²
项目组 / Team：崔彤 张润新 王康 陈希
业主 / Client：玉树州建设局

获首都第十八届城市规划建筑设计方案汇报展
优秀方案奖（2011 年）

北京太伟高尔夫休闲度假村
Beijing Taiwei Golf Club

地点 / Location：北京 / Beijing
设计 / Design：2002
建成 / Completion：2004
建筑面积 / Building area：35000m²
项目组 / Team：崔彤 赵正雄 曾荣 华夫荣
业主 / Client：北京太伟集团
摄影 / Photographer：杨超英 傅兴

鄂尔多斯 20+10 地块 T7 T4
Ordos 20+10, Porject T7 &T4

地点 / Location：鄂尔多斯 / Erdos
设计 / Design：2010
建成 / Completion：建设中 / under construction
建筑面积 / Building area：40000m²
项目组 / Team：崔彤 刘建平 苏东坡 王康 罗大坤 唐璐
业主 / Client：鄂尔多斯东胜区规划局

大连金石滩葡萄酒庄
Dalian Jinshitang Winery

地点 / Location：大连 / Dalian
设计 / Design：2012
建成 / Completion N/ A
建筑面积 / Building area：4760m²
项目组 / Team：崔彤 王一钧 刘建平 赵迎 刁建存 张润欣 曹炜
业主 / Client：大连金石葡萄酒庄有限公司

中国科技大学体育馆
Gymnasium of University of Science and Technology

地点 / Location：合肥 / Hefei
设计 / Design：2012
建成 / Completion：建设中 / under construction
建筑面积 / Building area：31000m²
项目组 / Team：崔彤 赵正雄 罗大坤 兰俊 彭相国 屠雁飞 杨向天
业主 / Client：中国科技大学

中国科学院研究生院新校区会堂及体育馆
Assembly Hall and Gymnasium of the New Campus, Graduate University of CAS

地点 / Location：北京 / Beijing
设计 / Design：2008
建成 / Completion：2014
建筑面积 / Building area：6000m²
项目组 / Team：崔彤 赵正雄 罗大坤 王一钧 吕僖
业主 / Client：中国科学院

中国科学院九区（奥运周边）项目
Section Nine of Chinese Academy of Sciences

地点 / Location：北京 / Beijing
设计 / Design：2012
建成 / Completion：N/ A
建筑面积 / Building area：391730m²
项目组 / Team：崔彤 白小菁 刘立森 王康
业主 / Client：中科院行管局

崔彤访谈

Interview

采访人 / 黄元炤
北京　2011.12.29

黄：1997 年清华大学建筑学硕士毕业后，您就被分配进入中国科学院建筑设计研究院工作，而您比较著名的作品是产生在 2002 年的中国科学院国家科学图书馆。我的体会在这个项目上，您尝试将中国传统文明在现代建筑中展现出来，一是以合院式的概念将几何体的一面挖空，二是暴露出许多巨大梁柱与桁架，犹如是召唤着斗栱的形制。另外您又用高大进退的墙板、柱廊与大台阶，创造出空间的层次感，企图将空间的尺度拉高与拉大而形成一种象征性的庄严气势，有一种巨大化、宏大化的设计倾向，能谈谈当时设计的意图吗？

崔："图书馆"其实是一些年来思考和感悟的结果，尽管有好多不遂心愿的地方，如内院尺度、材料选择、建造工艺等都有很多遗憾，但还是实现了自己原初的一些想法。这里最核心的问题是"心"的事，我所理解的中国人"中心"表面上看似是"无"和"空"，其实是充满期间弥漫于周围的"能量源"。就中国院子而言，与西方院子有着明显差异，你可以这样设想，一个方形封闭盒子，如果没有口，它是无所实用的，当上部中间有开口时，这个盒子就有了容器的可能。如果人在其中就可以生存，要注意，这就是中国空间的"原型"。就四合院而言，外边界是不开窗的，中间挖空的内院，首先是为了获得阳光和空气——即人类生存的最基本条件，也就是说"院"——"能量源"是中国空间中第一生存要素，这是客观的唯物的；接下来的才是基于天、地（自然）、人、物（建筑）为一体的宇宙观下的空间、形态和意境。图书馆作为"光"的容器，也是对阳光的接纳和遮蔽的平衡，这自然想到了合院空间中"能量源"。中间的"空"就是合院空间直接转译的结果。它给予了图书馆自然的光和风。

在某种意义上讲，这种"类合院"的空间或"U"字形的空间，也可以说是一种个人的"心理图形"。它可能来自于某种深刻记忆，比如刚才所说那个"书屋"式居所形态再叠加四合院的结构，居住在祖父留下的北京典型四合院里的体验也强化了这种"空间结构"的形成；图书馆的构成逻辑无外乎还是居所"书屋"的放大，两边书架放大成两侧的阅览空间，"书床"变为大台阶下的平台，"天花"变成了玻璃顶大厅和庭院的天空。

黄：您当时设计这个项目时，有参考国内外相关案例吗？

崔：比较遗憾的是，我们完成了方案之后，才参观和访问了法国、德国、英国的国家图书馆，当然收获不少。同时也更坚定我们原初的一些理念，比如当时在设计初期也有关于是整体式还是分散式之争，在看完法国国家图书馆的"四本书"、英国圣潘克拉斯火车站（St Pancras railway station）旁红砖砌筑的新不列颠国家图书馆以及德国的两个国家图书馆之后。好像更多是一种验证并确定我们方向的正确和一定的前瞻性，正如我们平常描述图书馆那样"集中式的便捷、分散式的环境优雅被融汇在一个理性平面中，从而确保阅览空间的采光和通风"。

黄：这个项目，在功能上已解决图书空间的问题，而在空间、形式语言与架构逻辑上，似乎让我观察到您有意来表达"中国性"的课题，将中国建筑中民族的情怀消化吸收于现代建筑之中，能谈谈这部分吗？

崔：是这样，这里有两个问题，首先需要明确作为国家科学图书馆"国家性"和"中国性"是绕不开的；第二，"中国性"设计并不是做完设计后再有意附加中国印迹。"中国性"设计是一种系统化设计，空间、形态和建构是难以剥离开的。我们之所以还沿袭西方理论那样分开在谈空间、形态和建构只是为了研究和表达需求。关于空间以上谈过了，与其说是还原一种合院空间的形态，还不如说是重新发现或重新解读合院，因为对"院"空间深入研究后我们发现它既不像有些人认识是剩下来的空间，也不是某些人把它奉为神秘莫测的精神场所。源于场地、源于功能，合院自然着陆其间，并以开放之势融通环境。至于中国建构的问题与院子是分不开的，既然院可称为"能量源"，那显然获取能量是通过院与房间之间界面空虚为目的，而木构成为了最好选择，长久以来中国建构发展区别于西方"砌筑式"建构体系，形成了几千年都不变的"架构"体系。我们为什么不问问这是"为什么"，至少，我们现在知道孤立地去谈建构或空间意义不大。

还有点要提的是中国的木构建筑也包括木塔，从建构角度或作为支撑体系来说不仅有其"现代性"和"透明性"，而且还具有高技派特征，我也常戏称中国曾有过"木构的高技"相信大家听说过以前在南方拆一座房子后，通过运河用船把拆下的房子在北京再重新搭建起来。仔细比较高技派的几大特征与中国木构建筑确有太多相似之处。因此，中国传统建筑与现代主义建筑已有天然的姻缘。

回到图书馆的建构问题，其实可以归结为"结构化的形式"和"透明性"的问题上来，中国木构建筑的"真实性"即结构逻辑与形式逻辑的对应关系，用"建构"来表达中国建筑尤为恰当。图书馆首先是将这种建造逻辑真实地再现出来。无论是类"穿斗式"的梁柱还是大跨度的桁架，还有受拉受压的杆件，都被转化为一种结构化的形式语言。

黄：这些“结构化形式”梁柱语言，仅仅在诠释中国性吗？它对图书馆的意义是什么？

崔：阅览空间的结构支撑体系“柱”被修正为“片柱”，目的是对光的有效利用和遮挡，并产生稳定的漫射光；西边柱廊的“透明性”昭示图书馆公共性和开放性，构成这个无顶无墙中央庭堂的界面；同时光和视线的穿越，确保南向阅览的采光和通风。

黄：您接着又设计中国科学院研究生院的教学楼，这个项目也是如您刚才所说的模式体现的吗？

崔：两个项目有相似之处，也存在着很多差异。

黄：也是中国性的架构语言？再谈谈您对中国建筑的研究。

崔：这两个项目的功能差异、场所差异以及体量差异，也造成两者出发点的不同。这个教学楼重点探讨的问题还不是架构语言。如果暂时抛开功能和环境问题，这里更多关注的应是空间类型，这一点与图书馆的相似在于“空间之间的空间”。简单地说两者都有一个“中心结构”和“时空次序”，无论是“C”形的“三合院”，还是两个平行体量所限定的“夹心院”或大厅，设计中都采用了基本的构成方式，如若干功能单元围绕一个中心空间，但与西方的教堂和巴西利卡不同的是中央的“透明”和“虚无”，并构成能量核心——一个充盈的“气”的流动，并不断地弥散，向某些方向延伸。图书馆的中心空间是向上和向西南的蔓延；而教学楼是向上和东西向的扩展。这些还是来自于中国传统建筑中建筑包围花园，或者建筑包容着一种“自然”，并以一种“好像”“与外隔绝”方式被引入的空间内部，这是才有意义。

与外界隔离的光和气是一个自然的世界，哪怕这个大厅的竹子、树也没有了。随着光的游戈和时间流逝，我们依然可以感知自然存在其中。还有人说这些大空间有些貌似教堂，我并不完全否认，无论是图书馆、博物馆还是科研教学楼，它们都有别于日常生活而成为一个特殊场所，不是一个世俗、生活情景化或者消遣娱乐的地方，我希望大家来这里找回自我、找回精神和信仰，于是宗教的氛围是存在的，但区别于西方教堂的神圣和向往，人文主义和自然主义的情趣让这些空间介于神圣与世俗之间。

黄：另外，这两个项目也呈现出一种宏大化、巨大化的表述，偏向象征性的设计语言。我为什么说象征，我说的象征不是“后现代主义”(Postmodernism)思潮的象征与符号，我说的象征指的是一种象征纪念性的设计路线，是“现代主义”(Modernism)的一条支流。在20世纪世界现代建筑史发展中，有一股象征主义或是纪念主义的流派，代表性建筑师有朱赛佩·特拉尼，他用柱梁的虚空构架，宣示出象征性的设计倾向。有路易斯·康，他将建筑的视点着眼于在巨大象征体下所创造出来的光影变化之美。还有安藤忠雄，他以清水混凝土营造出自然冥想的禅意巨大空间。从这三人的作品中创造出一条象征的设计路线，又有点理性，您的作品确实有这方面的倾向，您怎么去看您的作品与这些思潮、作品之间的关系？

崔：其实大与小是相对的，中国古代的城市无论是唐宋，还是元明清，城市的尺度都要比其他西方城市大得多，到现在这种趋势只增不减。就人口而言，北京市的2000多万人口能抵得上澳大利亚全国人口总数，而北京城的四环半范围就已经是新加坡国土面积那么大了，在这样尺度背景下，我们不得不思考关于城市及建筑空间尺度的问题，这好像好多人到了日本看丹下的东京市政厅，或看伍重的悉尼歌剧院都会感慨，怎么如此“小”，面对大山大水的北方，面对现代化的新“帝都”北京，我们必须重新建立一种与人关联的新尺度。另外一方面一种隐性的文化企图在东西方都长期存在：如神圣、伟大、纪念性、英雄主义往往相关联，也同时会产生积极和消极作用，比如大国、大都市、大教堂、大家、大学、大师、大智慧等。但同时中国还讲大象无形、大音稀声、大智若愚、大拙至美的辩证关系。

如果追求“假、大、空”或“高、大、全”，那叫为大而大，大而不当，甚至会产生乖张的荒诞景象。

对20世纪早期的象征主义建筑师，你刚才提到的朱赛佩·特拉尼不太熟知，好像是意大利理性主义代表人物，有点法西斯劲头。路易斯·康和安藤忠雄是我喜欢的建筑师，尤其是康，对他充满了崇敬，他可以成为我们这一代建筑师的导师；他不仅在巨大象征体下所创造出来的光和影，更多在于他用“可量度”空间所“滋生”出“不可量度”的宏大力量，并直抵你的内心。当然我们都知道“现代化”和“纪念性”是贯穿在康身上的两个激荡主题。这个过程中康试图将“建构性结构”与“纪念性形式”融合在一起，走出非同一般的道路。其中，古希腊和哥特的建筑思想，尤其是哥特的“结构化的形式”一度影响下，并让他迷恋“焊接钢管结构”构想，最后到对砌筑结构和框架结构清醒认知和得心应手。还让人敬佩的是他并不停留在结构理性主义上，而继续投向机电的热忱而创造吊顶夹层空间——而这一系列的探究，包括古希腊的神圣、哥特的光辉、结构的理性、机电的逻辑，汇聚在一起构成了一个宏大和神圣的纪念性篇章。

黄：路易斯·康和安藤忠雄的建筑是如何影响到您的？

崔：感受安藤忠雄的东西，可能还是源于东方人共同的理想和情趣。尽管安藤忠雄的房子都不大，但依然可以感受到一股自然的力量和圣洁的力量，这种少、无、禅、静的意象也形成一种“景生象外”的意境。我觉得安藤忠雄对大家的影响是心灵和精神层面上的，尤其是房子包裹着空灵的自然，是一种超越光与影之上“空间之间”的“膨胀感”和“扩散感”。

可以肯定的是现代主义中秩序或古典秩序，在康的身上表现出来的是重型结构形态和有层次的秩序，他也会成为我设计的营养。比如上两个作品明显地我要坚持“中心结构”和“时空秩序”。路易斯·康的思想已经熔炼了早期勒－杜克的结构理性主义和富勒的结构逻辑，也同时影响了我的设计，当然最为重要的影响应该是对建筑师“精神结构”的影响，正像是路易斯·康和安藤忠雄一样，我把设计当作一项精神活动，甚至有些宗教般狂热，你刚才所提到图书馆、开行、泰国文化中心、教学楼等作品在我看来也有着类似的宗教地位或崇高性，这样你可能不难理解我的作品中宏大的叙事结构和一定的纪念性。还有一些影响包括路易斯·康将对立的要素融合，并转化为一种积极因素，他的作品在稳定而厚重的对称图形中又融入了一种“哥特式”的智慧。但是和路易斯·康及安藤忠雄作品在心灵上的沟通可能还是源于共有的东方精神，一种在形而上的、对于“神”、“气”、“光”“风”以及对于“空间之间”、“天地之间”一种原始穿透力的痴迷。但同时我要说，我的设计出发点是中国的，或者说整体的秩序是中国的，而不是西方的。

黄：您谈到以柱与梁的结合产生传统中国性的架构语言，这似乎是一种折中语言的探索，从中国科学院国家科学图书馆到国家开发银行再到曼谷中心给我都是这样的感觉，很深的印象。而我所理解的折中分为中式折中和西式折中，我观察您更贴近于中式折中。是这样的，在19世纪鸦片战争后，许多现代化的思想、式样、材料、工法与技术经由租界所形成的渠道而来到中国，进而冲击到以木石系统为主的中国传统的营造思维，当时产生了许多以现代材料（水泥、混凝土、钢等）所建造出的民族复兴形式，并开启了一波大屋顶式的中华古典风潮，代表性建筑是1929年吕彦直设计的南京中山陵建成，当时中国近代建筑正摆荡在传统过渡到现代的折中状态。而我看您的作品似乎又看到那个年代里的一股折中的劲味，而折中实然是一项传统上的回复，或是在现代基础上的传统回复，一方面在传统基础上以片段的截取来重构在建筑上，或是同时都接受新的现代功能、结构与旧有形式的双重表现，是一项混血语法的展现，您对于折中的看法又是如何？

崔：我们先要提一下，你刚说“木石系统为主的中国传统营造”，这个说法还是要慎重一些，当然南京中山陵可能是个例外，因为砖石的确常用于陵墓建造中，以创造出一种永恒。至于以这种非典型场所和非典型类别的建筑和非典型的中国建构引领一个风潮，我还没有想明白。无论怎样，我的确认为中山陵是折中主义的，对于那个时代的人来说，这可能是历史中必然的，也算是一个突破，对传统认知和应用表现在如：①形式语言和符号语言；②空间语言和空间序列和群体组合语言；③“相地”语言和环境语言将建筑与环境相融等，对于过渡期的中国来说也算一种进步。在日本的帝冠式和我国台湾地区的一些建筑也都有这种印迹。可能在之后的如20世纪50年代和80年代又出现过类似的民族主义“形式”，建筑师尽管在这一时期创造很多不朽作品，但并没有明显发现中国建筑现代化中的跃迁，改革开放的最近20年来有了一些可喜的变化，作为一个中国建筑探索者，我还是认为折中建筑没有出路。

黄：我需要解释一下，吕彦直的中山陵完全是因为他在那个年代里的代表性与突破（材料的突破、形式的突破等），不代表所有建筑师都是这样遵循的，我所理解的折中是采取双重承认的态度，相对开放性的融合，当然有时也可间接的突破，所以在折中的大洪流下有着各种不同的倾向，在中国近代期间尤为明显，有的关注到材料倾向，有的关注到功能倾向，有的关注到新技术与传统历史的装饰结合的倾向，有的关注到城市中象征性的倾向等，所以折中是一种开放性的态度、思想。

崔：这可能是你对“折中主义”的一个“美好”的期待。我认为折中主义首先是一个时代的“叠加产物”，像是“过去时”＋“现在时”；同时折中主义是一种风格和形式上的“集仿形态”属形式主义。也可以说是一种物理上的“加法”，还处于一个中国建筑现代化中“初级阶段”。我所期待的中国建筑是一个交融的建筑，从根本上反对“模仿”和“拼凑”，也反对两者之间简单的叠加，“交融的建筑”如同两种东西放在一起发生化学反应一样，尝试将中国古典建筑与现代建筑作为基因或本源，通过溶解、渗透产生新的突变。

这首先要求我们去寻找中国建筑的基本原型，比如空间论中的“时空一体”；环境论中的“天人合一”；建构论中的“技艺合一”等，我们可以把它称为“一”的思维；然后它应该是“和”的建筑，强调异质要素的有机结合，体现在“和而不同”；最后它应该是“中”的建筑，但不是“折中”而是“执两用中”，它是在两极之间寻求新的动态平衡。

黄：您再谈谈国家开发银行，还是一个大屋顶，您在这里是如何体现“中国性”的？

崔：开行是有一个明显的玻璃屋顶和中国形式，这也难怪大家会误读，当然也有人认为艺术就在于“误读”，但开行对我来说并不是一个形式问题。它源于场地设计，关注“此时此地”以及“彼时彼地”，目标是建立一个平衡体系和透明体系。关于屋顶的事儿让我想起若干年前，我曾问过关肇业弟子张利的一个问题“为什么北大图书馆要用大屋顶”，他告诉我这是“应该”的和“唯一”的。

我8年前开始做开行设计，那是长安街最后的一块地，它北面是民族文化宫，它的南边只隔7米有6套文物级的四合院包括张学良的故居，再往南是大片的、相对完整的北京旧城。当时行长陈元有明确要求，提出八字方针“中国、现代、银行、生态”。并以此作为国际竞赛和招标的准则。起初我们的方案没有屋顶也没有基座，是一个抽象的只有中国建构逻辑的现代建筑，但在与甲方长时间的磨合中，他们希望并坚持在此基础“长出”屋顶和基座。虽然刚开始我不甘心，但在最后平衡过程中，我能接受这样结果是：除了这些“显性”的“类屋顶”之外，我们至少实现了中国建构逻辑的转译——一个八柱七间的巨构式的梁柱体系。

黄：这样的体系是一个现代银行所需要的吗？

崔：这个问题好，也是我们设计开始时问自己的问题，开行位于长安街西段，场所的敏感性并不是长安街上的几个建筑。

黄：民族文化宫是重要建筑，而且周边几个建筑也很重要，您认为这里最敏感的是什么？

崔：对，这些建筑都很重要！但它们也没有构成这个场所的“敏感性”，我幸好存有设计开始时——约2006年时这个地段的卫星截图，地段的南边是典型的旧城所呈现出的小尺度的院落肌理、胡同结构以及宜人的生活场景；而地段的北边则是新北京宏大的建筑群。纪念性或政治化的银行，超然尺度的长安街及空间形态。我们设计研究中发现了一个非常奇怪的问题。在长安街新建筑几乎都无视于这个亲切的、古雅的、有几何学精神的旧北京的存在！几乎无一例外的用一种实的、大的、封闭的形态将新、旧割裂成两个明显的区域。好像生怕这个“旧”的存在，或是影响了这个伟大北京的诞生！因此，我们认为场所的敏感性就是场所的“双重性”，它应该在大与小、新与旧、传统与现代、文化与文明之间建立一个平衡体系。

黄：那么说梁柱体系成了一个平衡体系。

崔：梁柱的架构体系本身还不是平衡体系，但它是形成这一体系的基础，中国的木构更接近于现代主义的框架系统，为创造自由平面、自由空间和自由立面创造了可能性。但最重要也最有价值的是中国木构逻辑所形成的结构清晰性和半透明性。因此，我们借用中国传统建造体系创造了一种非砌筑式、非封闭的体系以回应城市，具体的做法是从南边的传统肌理出发，探究四合院的空间形态和尺度，寻找一种可能性：是在城市尺度与旧城肌理、城市空间与院落空间、现代技术与传统技艺之间建立一种平衡体系。

前面说“敏感性”最终落在建筑“双重性”上，它存在于现代与传统两类城市形态的并置与冲突，设计的挑战以及主要“发力”都是如何在两种“异质”要素之间寻找平衡，它要求建筑具有“宽容度”和“沟通”能力。

黄：怎么理解“宽容度”和“沟通”能力？

崔：这是比喻，也就是说它一方面要与旧城对话，另一方面要与新北京对话的双重沟通能力和兼容性。恰好，架构体系的中国建筑就具有空透骨架的品质。我们把它理解为一个空透的网格，可以“吸附”、“吊挂”、“承载”任何物质，并且可以按照人为的方式“穿越”，这里封闭的墙或者盒子的概念完全被解体，光线、空气、景观以及人的行为按照设计目标可以自由穿越和贯通，如同是园林中“亭子”，提供了一个与周边沟通的框架。一个具有吸纳和释放的开放体系，它能够“连接”和“穿透”前、后、左、右、上、下，而成为新媒介。

黄："媒介"很有意思。

崔："媒介"是"之间"的状态，我试图寻找一个缓冲空间以消解长安街与旧城的冲突，并以此为契机，让这个"之间的空间"变成一个转换器，让南北两边最积极要素相互交融，变成一种能量体。这里包括几个问题，如何让阳光穿越建筑并照耀在开行北边的入口？如何让南边的七套四合院嵌入到建筑中来？如何让这个透明的体系装载更多的立体合院单元？

梁柱作为建构体的基本单元，在这个庞大的建筑中转化为巨柱和巨梁，传统木构体系演化为巨构体系，以便于形成现代银行所希望的大跨度、大空间，"巨柱"体系在保留传统"柱"支撑逻辑中变为"空间化的结构"柱筒，9m×9m的空间柱筒进一步转化为"束柱筒"，再以束柱筒为单元转化为"八柱七间"的"类木构"巨构秩序。

黄：这类"巨构"体系似乎常被现代银行采用，如香港汇丰银行、法兰克福商业银行，那您这里"巨构"是源于银行，还是源于中国建构。

崔："巨构"有助于大空间和特殊空间的形成，但另一个重要的背景是长安街，超然尺度诱发了这个想法，也成就了甲方"中国的"、"银行的"一种混杂纪念性的诉求。支撑这巨型"木构"也不偶然，中国古代由于用料的问题而产生的一捆束柱的做法也是一种建构手段，还有中国辽代留下来的应县木塔，从大逻辑上讲是筒中筒的概念。再细说，如八角形布局的每个角部也类似于三柱合一的巨构柱，但不同于应县木塔体系，也不同于香港汇丰银行体系，开行巨构是一个"空间化的结构"。柱子是"空心"的，是有实际使用功能办公单元。我们常开玩笑说，每一个巨柱都是一个"塔"或一个"高层建筑"，那么这个大房子就好像由16栋建筑合成，中间的"空"能让建筑的南、北最大限度的贯通。

因此，建构的目的并非是只为获得一个骨架，最终是要实现空间的建构。这一新体系区别于中国传统，也区别于现代主义的均质柱网和无柱空间的均质化。"巨构柱"的介入实现了从"支撑单元"向"空间单元"的转化。

黄：那屋顶也一定是空间逻辑和建构逻辑的结果了！

崔：中国古典的屋顶是建构逻辑形成的屋盖系统，屋顶下只有一个空间，屋顶自身并未形成独立空间。而开行的屋顶其实是"巨构梁"逐渐演化成为一个空间结构，它显然为顶层创造大空间"立了功"。最终我们看巨型梁柱所形成的"类抬梁"体系，转译为具有功能逻辑、空间逻辑的一种新透明体。它是一个大家能读懂的中国样式的建筑，但我们还深知它仍然保留大家"难懂"的，而我坚持下来的非"屋顶"的中国性。

黄：从中国科学院图书馆到国家开发银行再到几个文化中心和大使馆，如泰国的中国文化中心、法国的中国文化中心等，您的主线还是蛮清楚的，似乎每一件作品成形后都相当于一次理论与实践的总结，也导致这些作品都具有共同性。

崔：共同性是存在，但每个房子的场所不同，结果也就不同，有些重点于空间或空间之间，可能更多关注于"院"和"园林"空间对当代建筑的影响，这往往是一种看不着、摸不着的中国性，所以我想中国建筑，更多是"感觉性"或"听觉性"的，而非视觉性的。所以"空间的时间化"和"时间的空间化"可以创造出一种建筑的"视听感"，我正在尝试这种"时空一体"的可能。

还有相当一些建筑是讨论"建构"，如我们刚才聊的那些。中国建筑其实不是纯艺术，也不是纯技术，是"技艺合一"的，所以建构既不是形式问题，也不是构造问题，而这些建构问题又会与空间问题纠缠在一起，分不开；还有一些建筑是在关注自然、气候和建筑的关系，比较典型的是泰国文化中心，实际上它是一个源于场地、基于气候的"树屋"。

黄：是树上的"鸟居"，还是树状的建筑？

崔：东南亚地区的"树"很有意思，树冠很大而且稠密枝干好像还不是一个，是一组或一簇，在泰国寺庙和很多地方都能看到，在台湾也有这些树。

黄：我更关注树与文化中心的关联性以及和场地的关系。

崔：无论在泰国，还是中国，树木常认为是有象征意义的，其中一些甚至被人们看作神树，神树又与崇拜、创造、神话有关。在佛教国家中相传佛祖在菩提树下修炼成佛，菩提树又被认为是一颗无花果树的后代，因为这颗最原初的无花果树所栽的地方以及从此它的根系中生长出的树木所覆盖的地方，统称为"圣地"。位于泰国曼谷的中国文化中心，也可称为中泰文化交流的"圣地"，作为共同基因"佛祖"的敬仰，菩提树不仅是图腾，也是一个自然的建构体，可以挡雨、遮阳、通风，一个天然的庇护所。而且湿热地区中的菩提树（橡树）其枝上生有气根，垂入土中又成为另外的树干，所以，覆盖面积很大。因为菩提树象征着永生，气根象征着精神世界。我们的设计其实是学习这类被宗教化的自然秩序，在树干和气根形成的集合“树干”以及树冠的生长逻辑中做了一个“树屋”或者“仿生寺庙”。

黄：从气候到树、树到圣地，又到“寺庙”，从“寺庙”建筑再到“文化中心”是一个演进过程，我是否可以这样理解？

崔：这个逻辑是存在的。曼谷是被很多寺庙充斥的城市，什么样的现代文明都难以改变佛寺的统治地位，这种独有的“精神世界”也孕育出独有的城市文化和温文尔雅的泰国人。如果说“寺庙”是泰国民众的精神寄托的话，那么“文化中心”也可以成为“精神场所”，而“寺庙”的场所就变得有意义了。但同时，文化中心要去除寺庙的神秘而充满人性和自然的情趣，而这一转译又恰恰是中国儒道文化的人文主义和自然主义的精神。

具有中国传统精神的“架构”体系与“树构”的密切关联，使我们“建构”行为转化为“种植”活动。在这个特殊的场所中，气候、土地、树木是自然空间的要素，或者说是另类空间的要素，受自然律动的启发，而创造了一种新类型。建筑的自由释放和弥漫就像树木伸展和竭尽占领空间，最后密林般的柱子和横向密梁编织在一起，成为一个很茁壮的建筑。这个方法是从自然秩序发展出来的中国木构体系，又重新还原给自然。

黄：您其他的一些建筑是否也有类似的“自然主义”构想？

崔：自然可以是有形的，也可以是无形的，也可以形似或者神似。

像前面我们谈过自然可以成为“能量源”；当然自然也最容易产生一种意境；师法自然的设计也有助于我们创造出一种特殊的景象。

比如最近我们新完成的林业大学学研中心，实际上来源于林大特有的校园文化，如“知山知水”、“十年树木，百年树人”，同时是要提取树的这种生长或生发的意义，并且转化成一个精神上的“树”塔，借助于分形学上自相似性原则，进一步将“发芽”和“成长”抽象为繁衍式和渐变式的几何图形。

黄：您在最近完成的工艺美术馆和非物质文化遗产馆有一种水墨晕染的效果，形体也是渐变的，越来越模糊，同时又追求一种自然景象。

崔：我非常赞成“景象”这一表达。在工艺美术馆的设计中，虽然还在进行着建构的“研发”，并拓展和延伸到“编织”的技艺，完成了一个从砌筑到架构，又到编织出半透明的网纹的织物感建筑表达，但背后的意义，或者说建构意义只是为了表达一种自然景象，如：风、云、雨、雾的那种轻盈妙曼的意境，但建筑毕竟不是绘画，受到水墨晕染技法的启示，试图创造一个“立体的水墨画”。要做的是用现实的建构语言，重塑一种抽象自然，以梁柱构件作为空间“笔触”，表现自然中的光与色，努力让建筑成为一种对环境做出反应的有机体，建筑好像是源于自然秩序和自然景象而生成。

黄：您的作品很多，又多在“敏感”地段，虽然类型和风格也在不断变化，但我注意到有一种深藏于其中的不变性和一致性，而这个不变的“精神结构”似乎又在控制着您的建筑实践和思考，希望您再做一个小总结，谈谈对建筑的感悟。

崔：建筑是行与言、心与行、悟与心“二元中和”的产物。行、言、心、悟相互支撑、相互作用、相互转化，构成了一个动态平衡的开放体系。

行胜于言

建筑设计是行动主导下的图像建构，过度的建筑理论会产生副作用。“行胜于言”的重要性在于建筑设计应归于建筑实践的本源，让思想蕴含在物体之内，显现建筑真实存在的意义。设计作为“劳作”可以认知手工技艺如何决定机械技艺，又如何影响电脑科技。行胜于言在于动手。

心胜于行

建筑师有别于工匠在于学会思考“如何思考建筑”。“心胜于行”强调建筑师的“精神结构”对“身体结构”控制的对应关系，表现在心体合一、手脑共用，方可练就一双思考的手。肢体的感知、直觉的判断最终借助理性的智慧产生一种思辨的力量。

悟胜于心

设计过程是一种修炼的过程。设计中不断地积累、放弃、陈酿，终会有一个觉悟。“悟”源于实践之上，发展为超理性的感知系统，“觉悟”可以为孤思冥想，辗转心神之间，虽寄迹翰墨，以求景象万千。言、行、心、悟彼此氤氲化醇，最终获得对事物本质的认知。因此，不存在未经培训的先知先觉，设计便是“心思”和“觉悟”。

崔彤简介

崔彤，1962 年出生，清华大学建筑系毕业获建筑学硕士，国家优秀设计金奖获得者、全球华人青年建筑师奖获得者、享受国务院政府特殊津贴。现任中科院建筑设计研究院副院长、总建筑师、研究员；中国科学院研究生院建筑研究与设计中心主任，教授、博士生导师。中国建筑学会建筑师分会理事，外交部驻外机构工程建议咨询专家，中国美术家协会建筑艺术委员会委员，清华大学建筑学院设计导师，首都规划建设委员会专家咨询组专家，北京大学建筑学研究中心科技建筑研究所主任，《科研建筑设计规范》主编，《世界建筑》编委，《建筑技艺》编委，《a+a 中国建筑学会会刊》编委，《城市·环境·设计》编委，《城市·空间·设计》编委，《中国建筑文化遗产》编委。多次担任国家级及中科院重大项目的设计主持人和负责人，并获中国科学院创新先进个人。多个项目获奖，其中国家奖项 5 次，省部级奖项 20 余次。并应邀参加国内外学术交流和讲演，作品曾参展“从北京到伦敦：当代中国建筑展”（RIBA+UED）2012.4 英国伦敦；“建筑中国 100 作品”展 2012，德国曼海姆；“第六届圣保罗国际与设计双年展”2005，巴西圣保罗等。其作品和论文多次发表于国内知名专业杂志及出版物。2012 年入选当代中国百名建筑师。

崔彤·建筑工作室&建筑中心作为设计与教育的两个平台，长期致力于研究式设计与研究性教育，逐步形成了设计实践、研究、教育相结合的工作机制。